A Structured Elicitation Approach to Identify Technology-Based Challenges

With Application to Inform Force Planning for Technological Surprise

LAUREN A. MAYER, JON SCHMID, SYDNEY JEAN LITTERER, MARJORY S. BLUMENTHAL

Prepared for the Office of the Secretary of Defense
Approved for public release; distribution unlimited

NATIONAL DEFENSE RESEARCH INSTITUTE

For more information on this publication, visit **www.rand.org/t/RRA701-1**.

About RAND

The RAND Corporation is a research organization that develops solutions to public policy challenges to help make communities throughout the world safer and more secure, healthier and more prosperous. RAND is nonprofit, nonpartisan, and committed to the public interest. To learn more about RAND, visit www.rand.org.

Research Integrity

Our mission to help improve policy and decisionmaking through research and analysis is enabled through our core values of quality and objectivity and our unwavering commitment to the highest level of integrity and ethical behavior. To help ensure our research and analysis are rigorous, objective, and nonpartisan, we subject our research publications to a robust and exacting quality-assurance process; avoid both the appearance and reality of financial and other conflicts of interest through staff training, project screening, and a policy of mandatory disclosure; and pursue transparency in our research engagements through our commitment to the open publication of our research findings and recommendations, disclosure of the source of funding of published research, and policies to ensure intellectual independence. For more information, visit www.rand.org/about/principles.

RAND's publications do not necessarily reflect the opinions of its research clients and sponsors.

Preface

This report is part of a project on the nature and implications of ideas generated from across the RAND Corporation in response to an invitation to envision technology-based challenges that might emerge for the U.S. Joint Force around the year 2040. It supports a larger report, *Force Planning for Technological Surprise: An Argument to Imagine the Imaginable*, by providing greater methodological detail.

The project was sponsored by the Strategic Intelligence and Analysis Cell within the Office of the Under Secretary of Defense for Research and Engineering (OUSD[R&E]). It drew from open sources and unclassified discussions.

The research reported here was completed in December 2020 and underwent security review with the sponsor and the Defense Office of Prepublication and Security Review before public release.

This research was sponsored by the Office of the Secretary of Defense and conducted within the Acquisition and Technology Policy Center of the RAND National Security Research Division (NSRD), which operates the National Defense Research Institute (NDRI), a federally funded research and development center sponsored by the Office of the Secretary of Defense, the Joint Staff, the Unified Combatant Commands, the Navy, the Marine Corps, the defense agencies, and the defense intelligence enterprise.

For more information on the RAND Acquisition and Technology Policy Center, see www.rand.org/nsrd/atp or contact the director (contact information is provided on the webpage).

Contents

Figures and Tables

Summary

Objective

The Strategic Intelligence and Analysis Cell within the Office of the Under Secretary of Defense for Research and Engineering (OUSD[R&E]) asked RAND to draw on the diverse expertise of RAND staff to explore a wide range of possible challenges to Joint Force operations and the Department of Defense (DoD) that could, singly or in combination, compromise or even defeat the United States in a future conflict with a peer adversary. The research team approached this task in two ways. This report focuses on the development of a general approach, and its application, to nominate and evaluate such challenges using structured elicitation with internal RAND experts.

Approach

We developed a mixed-method approach to elicit and evaluate challenges (Figure S.1). Our approach begins with conducting and qualitatively analyzing a set of exploratory interviews with experts, which informs the next two stages of the approach. In the first, the expert framing of the problem is used to develop an open-ended survey to elicit nominations of challenges. In the second, it also heavily informs the development of a framework in which to evaluate challenges.

With this exploration complete, an open-ended survey is developed and used to elicit challenge nominations from a set of experts. The challenges are then assessed by a separate group of experts in the

Figure S.1
Expert Elicitation Approach

challenge evaluation. A close-ended survey is developed in which each nominated challenge is accompanied by evaluation questions determined by the criteria identified during qualitative analysis. The close-ended nature of the challenge evaluation form then allows for quantitative analysis of evaluation data within and between challenges.

Application of the Approach

Twenty-one RAND experts nominated a total of 40 challenges, which we consolidated into 20 challenges for evaluation (Table S.1). These challenges varied in terms of technologies, adversaries, and external factors, with potentially greater representation from challenges with technologies involving autonomy, machine learning or analysis of "big data," and the space domain.

Table S.1
Challenge Evaluation Clusters

Challenge Name	Short Description
Loss of anonymity	Facial, gait, and iris recognition lead to loss of anonymity and pseudonymity for servicemembers
Geoengineering	Geoengineered climate warfare
Minefields	Improved unmanned aerial systems (UAS), sensors, energy storage, and communications enable durable and effective minefields
Autonomous UAS	The proliferation of fully autonomous UAS and UAS swarms
An adversary dominates space	An adversary achieves space dominance
Anti-access	Artificial intelligence (AI), directed energy (DE), and counter-stealth create a new anti-access challenge
Virtual reality	The impact of virtual reality on training in the armed services
Spoofing vulnerability	Reliance on automated systems creates vulnerability to spoofing and data manipulation
Quantum computing	Quantum computing creates a world without encryption
Picosat swarms	Picosat swarms employed to attack high-value U.S. space assets
Sensing grid	Pervasive automated sensing grid alerts adversaries of U.S. troop movements and force levels
Bioweapons	Genetically engineered bioweapons
Adversary space communication	An adversary forms unbreakable quantum entanglement-based military space communication network
Adversary moon colony	An adversary establishes space superiority by creating first permanently manned colony on the moon
An adversary's tactics development	An adversary develops a military tactics development engine
Earthstrike capability	Adversary develops space-to-ground ("earthstrike") capability
Human adversaries	Overmatch from augmented human adversaries
DE weapons in space	Advanced DE and jamming/spoofing weapons used on U.S. space assets
Intelligent robots	Adversaries employ intelligent and merciless robots
Missiles and internet-of-things (IoT)	Ubiquitous IoT sensors enable targeting of U.S assets via hypersonic weapons

SOURCE: RAND analysis of expert nominator responses from n = 21 nominators.

Twenty-five RAND experts and 15 military fellows (active civilian or enlisted from across the DoD services who join RAND as visitors for one year) evaluated the 20 challenges across seven broad criteria that experts judged would classify a challenge as radically altering for the DoD: barriers to DoD identification and prevention; other barriers (e.g., feasibility of technology development or adaptation); mitigation difficulty; impact on operations; impact on national security; urgency of investment; and national defense priority.

This evaluation highlighted three clusters of challenges. A *top* cluster includes nine challenges that experts believed need urgent science and technology (S&T) investments, should be a major national defense priority, might have impacts with severe consequences, and overall were higher risk to the DoD and the United States. The *middle* cluster includes seven challenges that experts believed might have impacts with severe consequences but did not significantly need urgent S&T investment or to be a major defense national priority. A *bottom* cluster includes four challenges that experts did not believe had significance across the dimensions we measured.

Discussion

While a number of efforts to define and prioritize the most important or disruptive future technologies or technologically related risks have been attempted, the extent to which they were systematic, truly exploratory in nature, and analyzed using structured qualitative analysis is unclear. Our mixed-method approach rectifies many of these limitations, permitting a transparent, credible, and reliable means to identify and evaluate these types of challenges. Application of the generalized approach may be beneficial for other idea elicitation and evaluation tasks.

The clustering of challenges in our results should help to inform prioritization efforts within OUSD(R&E). One interpretation is to prioritize the challenges in the top cluster. Experts saw these as having an urgent need for DoD S&T resources and as potential national defense priorities. However, it is possible these evaluations were biased by current DoD efforts and that the challenges are already under consideration by OUSD(R&E). In that case, the challenges in the middle cluster may better receive priority. Possibly they are not as well known to our experts and therefore were not seen as important in terms of urgent S&T investments or national defense priority. Still, each of these challenges was rated as having significant impact on DoD or the United States if realized. If OUSD(R&E) intends to explore new, less intuitive challenges, the middle cluster should be prioritized. If instead its intent is to deepen knowledge in challenges potentially already being considered by DoD, the top cluster should be prioritized.

Acknowledgments

The research team is grateful to the many RAND colleagues who participated in the expert elicitation processes described in this brief report. We also appreciate the interest and support of the Strategic Intelligence and Analysis Cell within the Office of the Under Secretary of Defense for Research and Engineering at the Department of Defense, which supported the project for which this is one of three reports. Finally, we thank Philip Shapira and Laura Werber for their peer review of this report.

Abbreviations

AI	artificial intelligence
DE	directed energy
DoD	Department of Defense
IED	improvised explosive devices
IoT	internet-of-things
OUSD(R&E)	Office of the Under Secretary of Defense for Research and Engineering
S&T	science and technology
UAS	unmanned aerial systems

CHAPTER ONE

Introduction

The primary goal of RAND's 2040 Challenges project was to draw on the inherent and diverse expertise of RAND staff to explore a wide range of possible challenges to Joint Force operations and the Department of Defense (DoD) that could, singly or in combination, compromise or even defeat the United States in a future conflict with a peer adversary. The effort was meant to both inform the priorities of Office of the Under Secretary of Defense for Research and Engineering (OUSD[R&E]), as well as develop an approach to "crowdsourcing" ideas within RAND that could be applied to future policy questions.

Accomplishing this meant developing an approach and applying the approach to two related but differentiable tasks: *ideation*, which is the task of coming up with concepts for the challenges ("Sharks with lasers to threaten submarines!"), and *evaluation*, which is assessing the quality of each idea ("Laser energy propagates poorly underwater, and besides, the sharks would probably eat their trainers").

The research team conducted these tasks by developing two different approaches, each drawing from a different but potentially overlapping set of RAND participants. One of these approaches focused on using structured elicitation and evaluation of challenges from internal RAND researchers with expertise in relevant areas (i.e., RAND experts).[1] It is the focus of this brief report. The other approach combined an open solicitation of ideas with their vetting using a market

---

[1] This effort was determined to be NOT RESEARCH by RAND's Human Subjects Protection Committee (2019-0695).

mechanism, both open to all RAND staff. The outputs from both approaches were used to inform the main report of the project, *Force Planning for Technological Surprise.*

The remainder of this chapter includes a short review of existing approaches to technology risk prioritization, followed by an overview of the approach we developed. Chapters Two through Four provide a detailed description of our application of the approach and the corresponding results across three phases: exploratory interviews, challenge nomination, and challenge evaluation. The report concludes with a discussion of the results, as well as commentary about the value and future applicability of our developed approach.

Existing Approaches to Prioritizing Technology-Initiated Challenges

A number of organizations have attempted to define and prioritize the most important or disruptive future technologies or technologically related risks (e.g., Bidwell and MacDonald, 2018; Intel, 2018; Manyika et al., 2013; Miller, 2015; MIT Technology Review, 2020; NAE, 2017; WEF, 2019; WEF, 2020). These efforts use a variety of methods ranging from unstructured to structured, each with slightly different problem scopes. All involved recognized experts in their evaluation or assessment process. On the more unstructured end of the spectrum, for example, the National Academy of Engineering (NAE) identified 14 grand challenges for engineering by convening a panel with subsequent review by subject matter experts (NAE, 2017). Another effort by the Center for Strategic and International Studies, to identify drivers of the future security environment in 2045, was informed by various literature reviews, DoD documents and expert interviews (Miller, 2015). Other efforts chose more structured approaches. McKinsey and Company, for instance, identified 12 economically disruptive technologies by developing a candidate list of more than 100 technologies, informed by literature, press reports, and interviews, and assessed each candidate according to four criteria at least partially supported through rigorous analysis (Manyika et al., 2013). The World Economic Forum (WEF) has also led a number of relevant structured studies; in one focused on the top 30 global risks, a set of experts evaluated candidate risks according to their likelihood of occurring globally within the next ten years and their negative impact across countries and industries over the same time frame (WEF, 2020).

Given that existing approaches to elicit the most important or disruptive future technologies or technologically related risks from experts could have been adapted for the scope of our problem, the question arises: Why develop a new expert elicitation approach? The answer requires an understanding of how individuals make decisions. To elicit reliable, well-informed judgments from individuals that are consistent with their preferences, experiences, and values,[2] information and elicitations must be presented to them in a way that is relevant to their decisionmaking (Morgan et al., 2002). For example, close-ended (e.g., multiple choice or rating scale) survey questions that do not adopt the problem framing and language used by its respondents are less likely to capture accurate results of the question being asked. Instead, survey writers must understand the way their target audience thinks and talks about the problem to develop reliable survey questions (Converse and Presser, 1986). In the same way, when eliciting nominations or evaluations of challenges, we had to understand our experts' decision framing. The most effective means of gaining this level of understanding is through the systematic conduct and analysis of formative exploratory interviews. While a number of the efforts (presented above) did conduct interviews with experts, the extent to which they were systematic, truly exploratory in nature, and analyzed using structured qualitative analysis is unclear. Therefore, we determined the need to essentially start from the beginning by conducting our own formative research.

---

[2] Many individuals are willing to offer perceptions, preferences, or judgments as part of elicitations or surveys about topics for which they are uninformed, underinformed, or misinformed. This can be concerning, as such judgments may not reflect their true values, knowledge, and experiences and are often unstable, changing as an individual becomes more informed. As a result, the elicited judgments may be unreliable and of low quality (e.g., Price and Neijens, 1997).

Structured Elicitation Approach

When applied appropriately, expert elicitations can formalize and quantify judgments of individual experts about uncertain topics, for which quality data is lacking, to inform policy and resource allocation decisionmaking (e.g., Morgan, 2017). A well-structured expert elicitation approach provides a number of advantages over other less-structured, informal approaches or those that rely on nonexpert respondents. First, it can elicit well-informed judgments that are consistent with expert knowledge, experiences, and values. Second, it can mitigate cognitive heuristics (i.e., mental shortcuts) and biases to which all humans, even experts, are susceptible (e.g., Hastie and Dawes, 2009). Finally, it can provide transparency to the process and outputs, which can in turn lend credibility and confidence to the results.

However, to realize these advantages, the elicitation must be presented in a format that is relevant to the decisionmaking of the experts involved. Given the limited literature on structuring elicitations of technology-initiated challenges and the unique goal of this research, our approach began by performing exploratory research on how best to frame the problem. This involved first conducting and analyzing a set of open-ended interviews with experts. We then qualitatively analyzed these interviews to guide our formulation of questions to elicit both the nominations and evaluation of challenges. Through these interviews, we sought to understand experts' mental models (Morgan et al., 2002) or their internal conceptualization of a 2040 challenge. Our goal was to structure all subsequent interactions with experts using terminology and a framing that is relevant to, and aligned with, their mental model and related decisionmaking. Developing communications and elicitations that align with the mental model of the intended audience member has been shown to reduce the cognitive load of the participating decision-maker and produce judgments that are more reliable and consistent with the decisionmaker's internal preferences (Morgan et al., 2002).

Overview of the Approach

Figure 1.1 presents a high-level overview of our mixed-method approach to eliciting and evaluating challenges. We provide a brief overview of

Figure 1.1
Expert Elicitation Approach

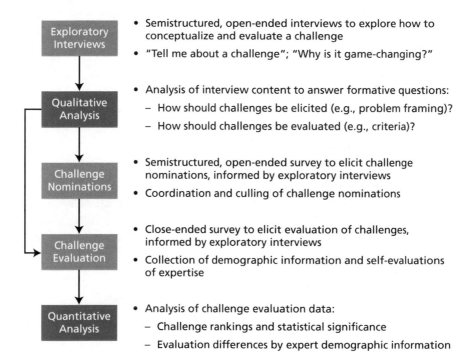

Exploratory Interviews
- Semistructured, open-ended interviews to explore how to conceptualize and evaluate a challenge
- "Tell me about a challenge"; "Why is it game-changing?"

Qualitative Analysis
- Analysis of interview content to answer formative questions:
  - How should challenges be elicited (e.g., problem framing)?
  - How should challenges be evaluated (e.g., criteria)?

Challenge Nominations
- Semistructured, open-ended survey to elicit challenge nominations, informed by exploratory interviews
- Coordination and culling of challenge nominations

Challenge Evaluation
- Close-ended survey to elicit evaluation of challenges, informed by exploratory interviews
- Collection of demographic information and self-evaluations of expertise

Quantitative Analysis
- Analysis of challenge evaluation data:
  - Challenge rankings and statistical significance
  - Evaluation differences by expert demographic information

the approach in this section, with further details for each stage presented in subsequent chapters on results.

As discussed previously, our approach begins with conducting exploratory research through a set of *exploratory interviews* with experts. Interviews are open-ended and use a "funnel design" (Morgan et al., 2002) in which beginning questions ask about broad concepts, with subsequent probing questions that are meant to elicit more specific detail. Use of a funnel design reduces the possibility of priming effects (i.e., interviewer presenting information that could bias participant responses).

Once interviews are complete, notes and/or transcriptions from recorded interviews are then assessed using a *qualitative analysis*. This analysis was informed by methods for thematic coding (Bernard et al., 2017; Braun and Clarke, 2006) in which emerging themes are identified, tagged in the text, and analyzed for patterns. The qualitative analysis of expert interviews is used to inform the next two stages of the elicitation approach. In the first, the expert framing of the problem is used to develop an open-ended survey to elicit nominations of challenges. In the second, it also heavily informs the development of a framework in which to evaluate challenges. Specifically, themes that emerge across more than one interview suggest criteria on which experts assess the value of presenting any specific challenge to the OUSD(R&E) sponsor. As themes are identified, they are categorized and culled into a tractable set of evaluation criteria.

With the exploratory research complete, an open-ended survey can be developed and used to elicit *challenge nominations* from a set of experts. Once nominated challenges are collected, the challenges can be reviewed for redundancy and applicability to cull them into a set of unique and relevant challenges for evaluation. The challenges are then assessed by a separate group of experts in the *challenge evaluation*. A close-ended survey is developed in which each nominated challenge is accompanied by evaluation questions determined by the criteria identified during qualitative analysis. Selected experts are then asked to respond to these questions as well as additional questions that are used to collect their demographic information and self-evaluation about their areas of topical expertise.

The close-ended nature of the challenge evaluation form then allows for *quantitative analysis* of evaluation data within and between challenges. Challenges can be compared and ranked using expert inputs to each criteria assessed, and the statistical significance of effects can be explored. Differences in expert assessments can also be explored based on demographic and expertise self-evaluation data.

Expert Sample Selection

As each phase of the elicitation approach involves recruiting experts, our method includes a structured approach for selecting three expert samples—a sample of *interviewees*, *nominators*, and *evaluators*. While the particular contribution of each sample to the overall approach varies, the three samples share a common objective: to use expert knowledge to inform the research project approach and its output. The objective of the samples has consequences for the type of sampling strategies employed. That is, this expert elicitation is not meant to obtain a statistically representative sample of the views of all RAND experts. Instead, the approach is used to develop an understanding of the range of appropriate expert judgments across RAND's population. To sample for this purpose, Morgan (2017, p. 253) notes that "care must be taken to include people who represent all the major perspectives and interpretations that exist within the community." Our approach, therefore, relies on purposive sampling, a nonprobabilistic method in which our project team selects experts based on which ones will be the most useful and/or representative (Babbie, 2015).[3] To perform this type of sampling, a list of experts is generated, with each expert being characterized based on the individual's background and topical area of expertise. From this list, experts are selected to ensure diversity across these key characteristics.

---

[3] Our application of the expert elicitation approach relied on inputs specifically and exclusively from RAND experts. As previously mentioned, our objective was to "crowdsource" ideas from expertise inherent to RAND. Sample selection for a broader set of experts may require further refinements to this approach.

CHAPTER TWO

Exploratory Interviews

Expert Interviewee Selection

For the Exploratory Interview phase, we sought to engage experts in broad-ranging, in-depth conversations regarding what constitutes a technology-initiated challenge to the Joint Force. To this end, we selected experts with expertise at the intersection of technological systems and Joint Force operations. To ensure diversity across military systems platforms and major operational domains, we selected one expert each from five major military operational domains: air, land, maritime, space, and information/cyberspace. Each selected expert was invited to participate. One expert declined and was replaced with an alternative choice with a similar set of expertise. Our resulting interviewee sample was comprised of three senior engineers, one senior physical scientist, and one senior management scientist, ranging in tenure at RAND between five and greater than 30 years[1] (mean = 15 years).[2]

Approach in Detail

We conducted 90-minute phone interviews with five RAND expert interviewees.[3] Prior to the interview, experts received a 30-minute

---

[1] Here, and in other parts of the report, the lack of precision to report specific years of experience was used to protect confidentiality of the expert.

[2] RAND tenure is not equivalent to years of experience. Many experts had previous careers in military services or other government organizations.

[3] One member of the team conducted all five interviews.

homework assignment to formulate ideas for three challenges for discussion: two meaningfully different 2040 challenges that they would describe as "radically altering" and one historical example challenge that they believed was game-changing at the time. We defined a "radically altering" challenge as follows:

> A technological, operational, organizational, and cultural challenge that has the potential to radically alter the strategic and operational environments the U.S. military may encounter by 2040. A challenge that might emerge because of the confluence of multiple technological developments. While these technologies can still be emerging, they should at least currently be at a stage in which they have been validated in a laboratory environment.

During the interview, we asked each expert to describe one 2040 challenge in detail and probed on a number of topics, including how they formulated the challenge and problems related to this formulation, the technologies involved, why they believed the challenge to be game-changing, and assumptions they used. Subsequent questions asked experts to discuss their problem framing for this exercise more generally. Next, we asked experts to compare the two 2040 challenges from their homework assignment in terms of which was more radically altering and their related rationale. Finally, we asked experts to describe their historical challenge, explain why they considered it to have been radically altering, and whether this historical calibration caused them to reevaluate any of their answers to previous interview questions. The homework assignment and full interview protocol are provided in Appendix A. We wrote interview notes using verbatim language. Audio-recordings of the interviews were used to supplement the notes, as necessary.

We identified common themes in the interviews using methods informed by thematic coding (e.g., Bernard et al., 2017; Braun and Clarke, 2006). Thematic coding allows qualitative text to be formulated as data for structured analysis and aids in summarizing key information from interview notes/transcripts. In its most rigorous form, the process involves systematically reviewing interview transcripts to develop a set of common themes and developing a "codebook" by creating a hierarchal structure of themes. Themes are then tagged in excerpts of text, which can then, for instance, be compared across interviews to explore patterns. Our qualitative analysis involved a streamlined approach. After a holistic review of the notes, we identified any concept discussed by experts in the five interviews that could inform two exploratory questions: How should challenges by elicited? How should challenges be evaluated?

To better understand how challenges should be elicited, we identified text that highlighted the decision context used by each expert, including how they framed the problem and the terminology they used. Questions and uncertainties that experts raised were cataloged. To determine how challenges should be evaluated, we identified concepts that could inform on how to evaluate a challenge in terms of its ability to be radically altering or game-changing.[4] Concepts across the five interviews were tabulated and sorted into common categories, which, in turn, were treated as resulting themes. Next, we reread each set of interview notes to determine whether a theme was mentioned. Themes being observed in two or more interviews were retained.[5]

Qualitative Analysis of Interviews: Results

Our qualitative analysis of the interviews revealed our expert interviewees' decision framing, as well as a number of themes, to help answer our two formative questions.

---

[4] Themes were identified using similar concepts to those raised in the interview protocol. That is, concepts had to answer one of the following questions to be considered a theme: What makes a challenge game-changing? If one challenge is more game-changing than another, what factors, characteristics, or qualities are being considered to make that evaluation?

[5] To ensure consistency and agreement on identified themes, two project team members individually identified themes in two of the interviews. Afterward, the team members compared themes and discussed those for which there was disagreement. Once agreement was reached, one team member took the lead identifying themes in the remaining three interviews. Areas of uncertainty for the remaining interviews were highlighted for review and discussion between the two team members. Other themes were reviewed and discussed periodically to ensure identified themes were consistently applied over the course of the exercise.

Challenge Formulation Decision Context

As discussed earlier, we began our interviews by asking experts to describe a 2040 challenge that they considered to be radically altering. Subsequently, we asked them to explain how they conceptualized the challenge and any difficulties they encountered. These probing questions elucidated a number of questions and uncertainties that experts were grappling with as they formulated a challenge. These uncertainties included the following:

- The technologies within scope
  - Are commercial or civilian technologies in scope?
  - Do technologies need to be emerging (or can challenges be initiated with technologies that are already widely used today)?
- Acceptable hypothetical pathways (e.g., external factors, assumptions) to challenge realization
  - Does (use of) the technology need to be the sole initiator of the challenge (or can it be in response to or in concert with some other initiating events)?
  - Does an adversary need to use the technologies (or can the United States or its allies be the user)?
  - Can a challenge emerge because of unintentional use or lack of use of a technology?
- Characterizations of the impact of the challenge
  - Are only technological impacts within scope (or could impacts be, for instance, operational, organizational, or cultural)?
  - Does the impact need to have a direct effect on the strategic or operational military environment (or could impacts be more indirect)?

This decision framing informed the design of our elicitation of challenge nominations from experts, as described in a later section.

Challenge Evaluation Themes

A large portion of the interviews involved the experts discussing why the challenges they described were considered radically altering. Our qualitative analysis of their answers to such questions elucidated a number of themes that could be used to characterize and differentiate the challenges. We identified 28 distinct themes mentioned by two or more of the experts that related to evaluating the challenges for the extent to which they were radically altering. These themes are presented in alphabetical order in Table 2.1. To illustrate the discussion surrounding these themes, we explain five in more detail below.

Ability to anticipate challenge All experts described the ability to anticipate a challenge as the "surprise factor." Experts approached this concept from two directions. First, they pointed to the difficulty of monitoring certain challenges or technologies that an adversary was secretly developing or progressing. These challenges, they suggested, could pose a more challenging problem for the United States than those whose development could be easily measured or predicted. Second, experts suggested that the United States could be "surprised" if it fails to recognize the potential of existing technologies to be used in new and devastating ways. For example, one expert discussed a historical example involving adversary use of improvised explosive devices (IEDs):

> That's what the IED did to us. Our vehicles were no longer sufficient—we couldn't even use a truck anymore . . . we didn't see it coming, and it was because of cognitive biases that we didn't. This wasn't a new invention that someone had created that we couldn't imagine being created. It more like something that was just so unsophisticated to us it wasn't worth our thought process before. And we paid the price, in billions and trillions of dollars, and lives lost, and the number of people with traumatic brain injuries. We paid the price of that mistake.

DoD ability to mitigate challenge When discussing how to decide which challenges should be considered radically altering, the experts we interviewed told us that they would consider the extent to which DoD was prepared to respond to the challenge. These discussions encompassed both DoD's capabilities and resources, as well as its authority to develop solutions. For example, one expert touched on the

Table 2.1
Themes Identified in Interviews

Theme	Description
Ability to anticipate challenge	DoD's ability to monitor or anticipate the development of a challenge or its enabling technologies and whether the challenge will catch DoD by surprise
Ability to respond to signs that a challenge is developing	DoD's ability to develop countermeasures prior to it occurring
Current challenge progress	How soon the challenge could be realized
Current preparations for challenge	U.S. or DoD current investment in preparing for this challenge
Current technology progress	Readiness level of the technologies enabling the challenge
DoD ability to mitigate challenge	DoD capabilities and authorities required to effectively mitigate the challenge; whether the organization of the U.S. government allows DoD to effectively mitigate the challenge
Ease of responding to challenge	How easily countermeasures for the challenge can be developed
Extent of challenge (among services)	How many services are affected by the challenge
Extent of challenge (in society)	How widespread and extensive the effects of the challenge are across U.S. society
Extent of challenge (scale)	The extent to which the challenge impact is widespread or disruptive
Extent of technology accessibility	How accessible the technologies enabling the challenge are to a wide variety of actors (both globally and at different scales)
Feasibility of technology development or adaptation to challenge	Whether the enabling technologies and supporting systems can be developed, operationalized, or adapted to military use
How tailored challenge is to U.S. vulnerabilities	To what degree the challenge or technologies are tailored to disrupt the U.S. way of operating or exploit seams in the U.S. national security enterprise
Impact on achieving national security objectives	Degree to which challenge threatens U.S. ability to achieve national security objectives

Table 2.1—Continued

Theme	Description
Impact on acquisition	Impact on acquisition process and outcomes (e.g., cost, schedule)
Impact on force design	Impact on how U.S. military organizes, trains, and equips
Impact on how U.S. operates	Impact on the way the DoD prefers to operate; impact on the options available for operations
Impact on military culture	Degree to which challenge changes U.S. military culture or is particularly difficult to manage because of U.S. military culture
Impact on relative operational advantage	Impact on the relative operational capabilities of the United States and its allies versus U.S. adversaries; impact on U.S. ability to win wars
Length of time needed to respond to challenge	How long will it will take to adapt to or counter the challenge, or for which the challenge's impact to be experienced
Preparedness for challenge	U.S. and DoD preparation to handle the development of the challenge
Relative ease of adoption (United States vs. adversary)	Difference in relative ability or willingness of the United States and its adversaries to develop and use the technologies enabling the challenge
Responsiveness to signs that a challenge is developing	How responsive the DoD is to signs that the challenge is developing
Risk of regret	Whether costs associated with failing to prepare for the challenge outweigh those to plan for the challenge, despite uncertainty over its occurence
Speed of technology progression	Speed and acceleration of development of technologies enabling the challenge
Technology growth potential	The potential of enabling technologies to become more widespread or useful in practical applications
Vulnerability of U.S. military	Overall vulnerability of the U.S. military or the technologies it depends on to the challenge and adversaries' abilities
Uncertainty over technology potential	Extent of uncertainty over whether the enabling technologies will perform as required to enable the challenge

SOURCE: RAND analysis of expert interviewee notes (n = 5 interviewees).

organizational challenges DoD might face when responding to hypothetical deepfake-enabled information operations against American society:

> This is something well outside the Department of Defense's lane, over in some combination of law enforcement and homeland security. We find ourselves handicapped by our worries on privacy and the right roles of the military and what the federal role is vis-à-vis the state vis-à-vis localities vis-à-vis private entities in policing, in some sense, this entire regime. I'm not sure how we're going to handle that.

This points to a concern also raised by multiple experts: that the organizational structure of the United States government might fragment authority in ways that constrain the ability of any one agency to find a solution to pressing challenges.

Impact on how the United States operates   Experts also consistently discussed challenges in terms of their impact on the way the United States conducts operations. Their thoughts were well summarized by one expert:

> I think of game-changing . . . as something that changes the entire way we fight. . . . To me, a game-changing technology is one that requires us to go back to pencil and paper and redesign every TTP [tactics, techniques, and procedures] and take apart every technology and put it back together in a different way again because it's all obsolete.

Unlike the previous theme of relative operational advantage, which addresses the effectiveness of the U.S. military relative to its adversaries, this theme considers whether a challenge would force the U.S. military to operate in new and potentially undesirable ways.

Preparedness for challenge   Experts also discussed how the lack of preparation for a challenge can have major consequences. This theme can be conceptualized as a precursor to the previously discussed theme of the DoD's ability to mitigate a challenge. Without preparations for a challenge, the DoD would likely be unable to mitigate it. The theme is also separate from that of anticipating a challenge. One can anticipate a challenge but still not prepare for it. One expert illustrated this idea particularly well using a historical example:

> The Royal Navy knew the Germans had submarines before 1914, but because they didn't really develop any effort to try to develop any kind of countermeasures to these things, early in the war—I think this was the first or second month of war—there were three big British cruisers sailing in the English Channel and one little German submarine sank all three of them in an afternoon . . . 1,500 British sailors were killed when those ships went down, and that shocked everybody. . . . No navy in the world—not just the British, but none of the other navies in the world—had done anything before the start of the war to say, "Okay, how do we counter this threat?"

Although the British Navy knew submarines were under development and did, in the end, develop countermeasures, its lack of preparation had devastating consequences.

Relative ease of adoption (United States vs. adversary) One final theme discussed by experts was the idea that a challenge might arise because a U.S. adversary was able to harness a set of technologies more effectively than the United States. One expert discussed this idea in the context of implementing brain-computer interfaces:

> There was some buzz the other year about soldiers and sailors using steroids in order to improve their physical performance. . . . There's an ongoing public debate about whether soldiers should take steroids to be better at what they do—and the standing position is, no, they shouldn't. . . . One of my concerns is that if this [the development of brain-computer interfaces] plays out as I suggest it might, some of our adversaries might not have any of these ethical qualms, or have less of them, so that they actually can get ahead. They will actually develop more warfighters who have direct implants, who can make quicker decisions.

In addition to ethical differences, experts pointed to differences in technology investments and governance structures as enablers to an adversary's ability to overmatch the United States in technological development.

CHAPTER THREE

Challenge Nominations

Expert Nominator Selection

During the Challenge Nomination phase, we sought to construct a sample of experts to submit challenge nominations. The primary objective of this phase was to produce a list of challenges that were expertly informed and diverse in terms of the initiating technological innovations. To this end, we employed a sampling strategy that sought representation from a broad set of knowledge domains. Specifically, we constructed the sample to ensure representation from the following knowledge areas: the air, land, sea, space, and information/cyber military domains of operation; biotechnology; energy technology; general (nonspecific) technological change; climate change; complex systems; critical infrastructure; special operations; and various regional areas of expertise.

To fill these domain categories with RAND experts, we solicited domain-specific nominations from the research team, the five expert interviewees from the Exploratory Interview phase, and RAND unit directors. This nomination process resulted in a list of 81 individuals, each of whom was categorized into a particular domain of expertise. We winnowed this list to a sample size of 40 based on the number of times that an individual was nominated. Of the 40 experts invited to participate in the challenge nominations, 21 returned completed forms. Table 3.1 provides a summary of the domain of expertise of expert nominators invited to participate, as well as those returning completed chal-

Table 3.1
Challenge Nomination Recruitment and Participation

Expertise Category	Expertise Domain	Experts Recruited	Experts Participating
Military domain	Air	4	1
	Land	4	1
	Maritime	4	3
	Space	4	2
	Information/Cyber	4	1
Other technology domains	Technology (general)	4	3
	Biotechnology	2	2
	Energy technology	2	1
Regional expertise	Various adversaries	5	1
Issue areas	Climate change	2	2
	Complex systems	2	2
	Critical infrastructure	2	1
	Special operations	2	1

SOURCE: RAND analysis of expert nominator expertise.

lenge nomination forms. Expert nominators' tenure at RAND ranged between one and greater than 30 years (mean = 10 years).[1]

Approach in Detail

To elicit challenge nominations from expert nominators, we developed a structured, open-ended survey. The survey was heavily informed by the challenge formulation decision context that emerged from our expert interviews in the previous phase (see Table 3.2). Once devel-

---

[1]   RAND tenure is not equivalent to years of experience. Many experts had previous careers in services or other government organizations.

Table 3.2
Challenge Nomination Form Items to Address Uncertainties Raised During Exploratory Interviews

Uncertainties Raised	Response (as presented in challenge nomination form)
Are commercial or civilian technologies in scope? Do technologies need to be emerging (or can challenges be initiated with technologies that are already widely used today)?	The challenge you nominate may have the following characteristics: Involve technologies that may be: Currently used or intended for use in commercial, civilian, and/or military situations.
Does (use of) the technology need to be the sole initiator of the challenge (or can it be in response to or in concert with some other initiating events)?	The challenges you nominate must have the following characteristics: Be at least partially initiated by two or more technological developments that come together in some way.
Does an adversary need to use the technologies (or can the United States or its allies be the user)?	The challenge you nominate may have the following characteristics: Involve technologies that may be potentially adopted by the United States, its allies and/or its adversaries.
Can a challenge emerge because of unintentional use or lack of use of a technology?	The challenge you nominate may have the following characteristic: Occur through intentional or unintentional use of the technologies or through the lack of use of those technologies.
Are only technological impacts within scope (or could impacts be, for instance, operational, organizational, or cultural)?	The challenges you nominate must have the following characteristic: Generate impact that can be technological, operational, organizational, and/or cultural.
Does the impact need to have a direct effect on the strategic or operational military environment (or could impacts be more indirect)?	The challenges you nominate must have the following characteristic: Have the potential to radically alter the strategic and/or operational environment the U.S. military may encounter by 2040.

SOURCE: RAND analysis of expert interviewee notes (n = 5 interviewees).

oped, the survey was iteratively refined based on pilot tests with a few members of the intended audience. Our resulting challenge nomination form asked expert nominators to nominate two challenges—one within and one outside their primary area of expertise. This method sought to ensure representation across the domains of knowledge while still allowing for nominations to come from domains other than those listed in Table 3.1. Experts were provided with clear guidelines to appropriately scope the challenges to the OUSD(R&E) sponsor's area of interest. For each challenge, experts were asked to provide a brief summary of the challenge, followed by descriptions of the technologies involved, external factors and assumptions involved, and the impact and importance to the OUSD(R&E) sponsor. The challenge nomination form is provided in Appendix B.

With nominated challenges collected, we reviewed the large set for redundancy and applicability to the OUSD(R&E) scope. We combined similar challenges and removed those that were out of scope to achieve a set of unique and relevant challenges for evaluation.

Challenge Nomination Results

The 21 expert nominators submitted a total of 40 challenges.[2] Our culling and consolidation of these challenges resulted in a total of 20 challenges for evaluation, with eight removed for being out of scope and the remaining challenges combined to eliminate redundancy. The most common reason for removal of a challenge was the lack of direct impact on the Joint Force.[3] For example, one respondent proposed a challenge involving large-scale exit from the electrical grid driven by improvements in energy production and storage technology performance. While this challenge is initiated by technological change and would likely have broad economic impact, its effect on the Joint Force is likely to be indirect. Challenges with similar themes and/or technologies were also common. For example, three nominations focused on the proliferation of unmanned aerial systems (UAS) and UAS swarms. These were combined into a single challenge called "the proliferation of fully autonomous UAS and UAS swarms" that incorporated aspects of each nomination. The resulting 20 challenges for evaluation are shown in Table 3.3, with full descriptions of selected challenges provided in Appendix C.

---

[2] Two of the 40 experts submitted only one challenge.

[3] This scoping provision was not included in the challenge nomination form and was determined based on sponsor interest after the form had been sent to experts.

Table 3.3
Consolidated Challenges from Expert Nominations

Challenge Name	Short Description
Loss of anonymity	Facial, gait, and iris recognition lead to loss of anonymity and pseudonymity for servicemembers
Geoengineering	Geoengineered climate warfare
Minefields	Improved UAS, sensors, energy storage, and communications enable durable and effective minefields
Autonomous UAS	The proliferation of fully autonomous UAS and UAS swarms
Adversary dominates space	An adversary achieves space dominance
Anti-access	Artificial intelligence (AI), directed energy (DE), and counter-stealth create a new anti-access challenge
Virtual reality	The impact of virtual reality on training in the armed services
Spoofing vulnerability	Reliance on automated systems creates vulnerability to spoofing and data manipulation
Quantum computing	Quantum computing creates a world without encryption
Picosat swarms	Picosat swarms employed to attack high-value U.S. space assets
Sensing grid	Pervasive automated sensing grid alerts adversaries of U.S. troop movements and force levels
Bioweapons	Genetically engineered bioweapons
Adversary space communication	An adversary forms unbreakable quantum entanglement-based military space communication network
Adversary moon colony	An adversary establishes space superiority by creating first permanently manned colony on the moon
Adversary tactics development	An adversary develops a military tactics development engine
Earthstrike capability	Adversary develops space-to-ground ("earthstrike") capability
Human adversaries	Overmatch from augmented human adversaries
DE weapons in space	Advanced DE and jamming/spoofing weapons used on U.S. space assets
Intelligent robots	Adversaries employ intelligent and merciless robots
Missiles and internet-of-things (IoT)	Ubiquitous IoTs sensors enable targeting of U.S assets via hypersonic weapons

SOURCE: RAND analysis of expert nominator responses from n = 21 nominators.

CHAPTER FOUR

Challenge Evaluation

Expert Evaluator Selection

To select expert evaluators for the Challenge Evaluation phase of our expert elicitation, we began with the existing list of 81 experts developed for the Challenge Nomination phase. Given that the two phases require the same functional role of expertise (i.e., to inform the process of producing a diverse and informed set of challenges), the domains of knowledge for which representation was sought was also identical. From this list of 81 experts, we excluded experts who participated in the first two phases of the approach, leaving a total of 50.

In addition to RAND experts, we added a second group to our sample comprised of RAND military fellows who join RAND as visitors for one year to gain research experience. Fellows are active civilian or enlisted from across the DoD services, with rank or rank-equivalents usually between major and lieutenant colonel. At the time of recruitment, RAND had 21 active military fellows. By including these military fellows in our sample, we sought to provide real-world military perspectives to evaluating challenges.

From our list of 81 RAND experts and military fellows, we selected an initial subset of 35 and 15 expert evaluators, respectively. The subsets were selected to diversify across domain expertise (for RAND experts) and military service (for fellows). An additional 15 RAND experts and six military fellows were chosen for reserve in case participation from the initial 50 experts was insufficient. In total, we invited 45 RAND experts and 15 military fellows to participate in the challenge evaluation, with 25 and 10 individuals, respectively, completing participation.

The 25 RAND expert evaluators ranged in tenure from one year to greater than 30 years (mean = 9 years).[1]

Approach in Detail

As discussed in Chapter 1 (Exploratory Interviews), we identified 28 themes that expert interviewees believed should influence how challenges are evaluated in terms of being radically altering. We used these themes to inform the development of a close-ended survey to elicit expert evaluator assessment of the nominated challenges across a set of evaluation criteria. In the survey, challenges were presented using a standard template: challenge name, summary, technologies involved, external factors or assumptions involved, and impact and importance to OUSD(R&E). Expert evaluators were asked to evaluate each challenge against seven evaluation statements that were directly informed by the interview themes (the translation of themes to evaluation statements is discussed in detail in the next section), using a scale from strongly disagree (= 1) to strongly agree (= 7).[2] Higher ratings implied that the challenge was more radically altering. The evaluation statements and their associated summary-level criterion name are shown in Table 4.1. Additional survey questions gathered demographic information and the expert evaluators' self-evaluation of their expertise across a set of topics. We pilot-tested the survey on members of the intended audience and subsequently iteratively refined its contents. We created two evaluation surveys, in which the ordering of challenges in the second survey was inverted from those in the first, and randomly assigned experts to each survey. An excerpt of the challenge evaluation form is provided in Appendix B.

---

[1] RAND tenure is not equivalent to years of experience. Many experts had previous careers in services or other government organizations.

[2] Experts had the additional options to respond that the statement was not applicable or that they did not know.

Table 4.1
Challenge Evaluation Criteria and Associated Survey Evaluation Statements

Number	Criteria	Statement (on a scale from strongly disagree [=1] to strongly agree [=7])
1	Barriers to DoD detection and prevention	DoD is unlikely to detect this challenge prior to it being realized or to prevent it from being realized even if it is detected (e.g., because of a lack of ability or willingness, or other barriers)
2	Other barriers	Other than barriers to DoD detection and prevention, there are very few barriers that could keep this challenge from being realized. Barriers may include, for example, lack of technology readiness or adversary ability/intent to adopt technology, forces that change external factors, and U.S. institutional or cultural incentives.
3	Mitigation difficulty	If this challenge is realized, it will be difficult for DoD to mitigate its impact (e.g., through authorities, capability, knowledge, costs, effort, or time)
4	Impact on operations	Left unmitigated, this challenge will severely limit (directly or indirectly) DoD's options for performing operations
5	Impact on national security	Left unmitigated, this challenge will severely negatively impact (directly or indirectly) U.S. national security (e.g., through changes to economy, culture, politics, military operations)
6	Urgency of science and technology (S&T) investment	If DoD does not invest soon, specifically in appropriate S&T areas, the opportunity to reduce the risks of this challenge being realized will have passed
7	National defense priority	Preparing for this challenge should be a major national defense priority

SOURCE: RAND analysis of expert interviewee notes (n = 5 interviewees).

Development of Challenge Evaluation Statements

To develop the evaluation statements, we began with a holistic review of the 28 themes and then sorted the themes into like groups. Given that expert evaluators were required to evaluate 20 challenges against the evaluation statements, our objective was to find a balance between keeping expert inputs to a manageable number and ensuring that the nuance in themes was preserved. The sorting resulted in five groups, in which the overarching concept of each group was developed into the first five evaluation statements shown in Table 4.2. We subsequently added two more evaluation statements to allow for a holistic view of challenge priority, as well as to encompass a few remaining themes that were not easily sorted into our original five groups. Pilot tests allowed for refinements to these statements along the way. A mapping of the 28 themes to the seven evaluation criteria/statements is shown in Table 4.2. Some themes clearly mapped to multiple criteria, and in those cases, we list them more than once. However, for brevity, the mapping shown is not fully exhaustive—themes that apply to criteria to a lesser extent or indirectly are not included.

Table 4.2
Mapping of Themes to Evaluation Criteria

Evaluation Criteria	Theme
Barriers to DoD detection and prevention	Ability to anticipate challenge
	Vulnerability of U.S. military
	Ability to respond to signs that a challenge is developing
	Responsiveness to signs that a challenge is developing
Other barriers	Relative ease of adoption (United States vs. adversary)
	Extent of technology accessibility
	Feasibility of technology development or adaptation to challenge
	Current technology progress
	Technology growth potential
	Uncertainty over technology potential
Mitigation difficulty	DoD ability to mitigate challenge
	Preparedness for challenge
	Ease of responding to challenge
	Length of time needed to respond to challenge
	Vulnerability of U.S. military
	Current preparations for challenge

Table 4.2—Continued

Evaluation Criteria	Theme
Impact on operations	Impact on relative operational advantage
	Impact on how U.S. operates
	How tailored challenge is to U.S. vulnerabilities
	Impact on acquisition
	Impact on force design
Impact on national security	Extent of challenge (scale)
	Extent of challenge (among services)
	Extent of challenge (in society)
	Impact on achieving national security objectives
	Impact on military culture
Urgency of S&T investment	Risk of regret
	Current challenge progress
	Current preparations for challenge
	Current technology progress
	Speed of technology progression
National defense priority	Impact on relative operational advantage
	Extent of challenge (scale)
	Risk of regret
	Extent of challenge (in society)
	Impact on achieving national security objectives

SOURCE: RAND analysis of expert interviewee notes (n = 5 interviewees).

Challenge Evaluation Results

Thirty-five expert evaluators evaluated the 20 nominated challenges across the seven evaluation criteria. We performed numerous quantitative analyses of the challenge evaluation data to inform discussions about which challenges were considered more radically altering. First, we created three composite measures, using the first five evaluation statements shown in Table 4.1.

- A *likelihood* composite calculated as the mean of the first two criteria in Table 4.1 (Barriers to DoD detection and prevention, Other barriers)[3]
- A *consequence* composite calculated as the mean of criteria three through five in Table 4.1 (Mitigation difficulty, Impact on operations, Impact on national security)
- A *risk* composite calculated as the mean of the first five criteria in Table 4.1 (Barriers to DoD detection and prevention, Other barriers, Mitigation difficulty, Impact on operations, Impact on national security).

Composite measures were calculated for each expert evaluator.[4] These three composite measures and the two remaining criteria (Urgency of S&T investment, National defense priority) formed the five evaluation criteria we used in our quantitative analyses.

Mean and standard deviation values were then calculated across the 35 expert evaluators for each criteria, across the 20 challenges. We finally calculated one-sample t-tests for each mean criterion against the scale midpoint of 4, to test whether the mean was significantly greater than this midpoint. Given that our evaluation scale implies an association between higher ratings and higher-priority challenges, this t-test was meant to suggest potential thresholds for which challenges to consider as radically altering. While the objective of this report is to focus on the methodology, the specific challenges have been anonymized when reporting rating results.

---

[3] The use of the term *likelihood* for this composite is not meant to suggest experts' true evaluation of the likelihood of the challenge. As we noted earlier in this chapter, our exploratory interviews suggested that when experts are asked to estimate likelihoods for these challenges, they likely provide erroneous estimates because this type of estimation would be extremely difficult, if not impossible. Instead, the term is used to suggest that a challenge may be more likely if it has fewer barriers to being realized. This is discussed further in a subsequent section of the chapter.

[4] If an expert omitted an answer or answered "don't know" or "not applicable" on any individual evaluation statement, composite measures using that criteria were also omitted.

Mean Ratings Results

Figures 4.1 and 4.2 present scatterplots of the mean challenge ratings for Urgency of S&T investment and National defense priority, and Likelihood and Consequence, respectively. Each figure additionally indicates whether the mean rating is significantly different ($p < 0.05$) than the scale midpoint. Given that the Risk composite is a combination of Likelihood and Consequence, Figure 4.2 also shows whether mean ratings for Risk are significantly different from the scale midpoint.

A holistic review of the scatterplot of mean ratings for Urgency of S&T investment and National defense priority, as shown in Figure 4.1,

Figure 4.1
Mean Ratings and Significance of Urgency of Science and Technology Investment and National Defense Priority

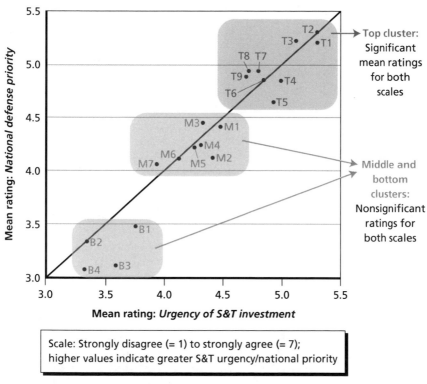

Scale: Strongly disagree (= 1) to strongly agree (= 7);
higher values indicate greater S&T urgency/national priority

SOURCE: RAND analysis of expert evaluator survey responses (n = 35).
NOTE: Diagonal line shown on graph is a y = x reference line.

Figure 4.2
Mean Ratings and Significance of Likelihood, Consequence, and Risk

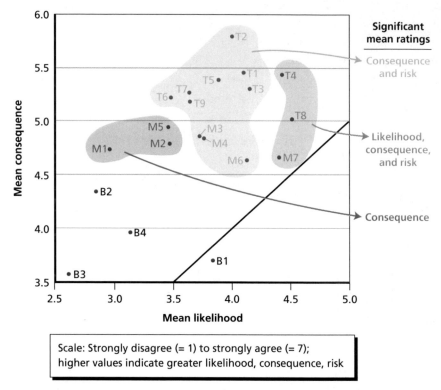

SOURCE: RAND analysis of expert evaluator survey responses (n = 35).

NOTES: Diagonal line shown on graph is a y = x reference line. Challenge labels in green represent challenges with mean ratings significantly above the scale midpoint for the consequence and risk criteria. Labels in orange are for the likelihood, consequence, and risk criteria; labels in blue are for the consequence criterion.

suggests that mean ratings of the challenges between the two measures were roughly similar (i.e., follow closely to the y = x line). This suggests that, generally, expert evaluators believed that challenges of more urgency for S&T investment were also of higher national defense priority. We also see roughly three clusters of challenges emerging in the scatterplot. The top cluster includes nine challenges rated highest on average on these two criteria. Mean ratings for each of these challenges were significantly greater than the scale midpoint ($p < 0.05$). A middle and bottom cluster comprise seven and four challenges, respectively. All challenges in these two clusters were not significantly greater than the scale midpoint for either criterion.

Reviewing the scatterplot of mean challenge ratings for Likelihood and Consequence measures, as shown in Figure 4.2, shows little to no clustering of challenges, as well as less symmetric ratings between the two measures (i.e., mean ratings do not align to the y = x line). Instead, expert evaluators rated all but one challenge as having relatively higher consequence when compared with likelihood. In fact, all but four challenges (the same four challenges in the bottom cluster in Figure 4.2) had mean Consequence ratings significantly greater from the scale midpoint ($p < 0.05$). Conversely, only three challenges had significantly greater mean Likelihood ratings when compared with the scale midpoint. Thirteen challenges had significantly greater Risk ratings.

Figure 4.3 provides the mean ratings for the 16 challenges in the top and middle challenge clusters. Each challenge was rated as significantly greater than the scale midpoint ($p < 0.05$) on at least one of our five evaluation criteria, with the nine challenges in the top cluster showing significance on at least four of the five criteria.[5] Two challenges—T4 and T8—showed significant ratings on all five criteria. Only one other challenge also had a significant Likelihood rating—that of M7.

While significance of mean ratings provides some indication of expert agreement on challenge ratings, a holistic review of their standard deviations (as shown in Appendix D) further suggests that expert evaluators were in less agreement on challenge ratings for the middle and bottom challenge clusters. That is, there is generally less dispersion for the Urgency of S&T investment, National defense priority, and Consequence criteria for those challenges in the top cluster. For example, standard deviations for the T3 and the T1 challenge were smaller than at least 17 of the other challenges for all three criteria. No similar pattern was found for the remaining 12 challenges on these criteria.

---

[5] The four challenges not shown were not rated as significantly greater than the scale midpoint on any of our five measures.

Figure 4.3
Mean Expert Ratings for the Top 16 Challenges

Challenge	Mean Rating				
	S&T Urgency	National Priority	Likelihood	Consequence	Risk
T1	5.30	5.20	4.11	5.44	4.90
T2	5.30	5.30	4.02	5.78	5.09
T3	5.13	5.22	4.17	5.29	4.86
T4	5.00	4.85	4.44	5.42	4.98
T5	4.93	4.65	3.90	5.38	4.77
T6	4.85	4.85	3.49	5.21	4.52
T7	4.81	4.94	3.65	5.26	4.60
T8	4.73	4.94	4.53	5.00	4.81
T9	4.71	4.88	3.66	5.18	4.59
M1	4.48	4.41	2.97	4.73	4.07
M2	4.41	4.13	3.48	4.78	4.22
M3	4.33	4.45	3.73	4.85	4.39
M4	4.32	4.24	3.77	4.83	4.45
M5	4.26	4.22	3.47	4.94	4.30
M6	4.13	4.12	4.14	4.63	4.41
M7	3.94	4.06	4.41	4.66	4.51

Top Cluster (T1–T9); Middle Cluster (M1–M7)

SOURCE: RAND analysis of expert evaluator survey responses (n = 35).
NOTE: Means significantly greater than scale midpoint using t-test: ■ $p<0.001$
■ $p<0.01$ ■ $p<0.05$.
Scale: Strongly disagree (=1) to strongly agree (=7); higher values indicate greater level of measure.

Furthermore, no patterns around dispersion are evident for the Likelihood and Risk ratings for the 20 challenges.

Challenges in Figure 4.3 are presented in descending order of mean ratings on the Urgency of S&T investment criteria. Reviewing the means for the other evaluation criteria shows that a rank ordering of the challenges by each criterion varies. For ease of visualizing these rankings, Table 4.3 presents a color-coded version of challenge rankings for each criterion. While rankings for Urgency of S&T investment, National defense priority, Consequence, and Risk are closely aligned, Likelihood rankings differ, with four of the top ten challenges ranked by likelihood falling in the bottom half of the table. This difference in Likelihood ratings is also evident in calculations of the correlations of expert ratings of individual challenges between criteria. For example, expert Consequence ratings are significantly correlated with both Urgency of S&T investment and National defense priority ratings for all 20 challenges (using Pearson's r test, $p < 0.05$—data not shown). However, correlations between expert ratings on Likelihood with these two criteria were only significant for ten and 12 challenges, respectively.

Population Differences in Mean Ratings

To assess whether mean ratings systematically differed for certain populations within our expert evaluator sample, we performed a large number of two-sample t-tests on mean challenge ratings for three criteria: Urgency of S&T investment, National defense priority, and the Risk composite. We first compared these mean ratings for each of the 20 challenges between experts who received the two different evaluation surveys discussed previously, in which the ordering of challenges in one (n = 18 experts) was inverted from the other (n = 17 experts). Using this test can help determine the presence of response-order effects, a cognitive effect in which survey respondents may answer questions differently based on the order they are presented (Schuman and Presser, 1996). Of the 60 two-sample t-tests, we found no significant differences in mean challenge ratings between these two groups ($p > 0.05$ for all), suggesting that no ordering effects were evident.

We further compared the three mean evaluation ratings for the 20 challenges between our two groups of expert evaluators: RAND experts and military fellows. Of the 60 two-sample t-tests, 58 of the tests showed no significant difference in ratings between these two populations ($p > 0.05$ for all). The two tests that showed significant

Table 4.3
Ranking of Challenges by Evaluation Ratings

Challenge	Urgency in S&T Investment	National Defense Priority	Likelihood	Consequence	Risk
T1	1	3	6	2	3
T2	2	1	7	1	1
T3	3	2	4	5	4
T4	4	8	2	3	2
T5	5	9	8	4	6
T6	6	7	14	7	9
T7	7	4	13	6	7
T8	8	4	1	9	5
T9	9	6	12	8	8
M1	10	11	18	14	16
M2	11	14	15	13	15
M3	12	10	11	11	13
M4	13	12	10	12	11
M5	14	13	16	10	14
M6	15	15	5	16	12
M7	16	16	3	15	10
B1	17	17	9	19	18
B3	18	19	20	20	20
B2	19	18	19	17	17
B4	20	20	17	18	19

SOURCE: RAND analysis of expert evaluator survey responses (n = 35).

NOTE: Colors provide a visual aid for rankings with red representing lowest rank (highest rating), yellow representing middle ranks, and green representing high ranks (lowest ratings).

differences ($p < 0.05$) were for the Urgency of S&T investment criteria of the M7 challenge and the B1 challenge. In the case of the M7 challenge, RAND experts rated the Urgency of S&T investment significantly higher than military fellows. In the case of the B1 challenge, military fellows rated the Urgency of S&T investment significantly higher than RAND experts. Given that our total sample size was 35 experts, however, it is possible that we did not have sufficient statistical power to assess potential systematic differences that could be occurring.

CHAPTER FIVE

Discussion

Challenge Evaluation

Our 35 expert evaluators evaluated 20 challenges, nominated by their peers, across seven dimensions of the challenge. These dimensions were designed to indicate a prioritization of challenges in terms of being radically altering and were determined based on formative interviews with RAND expert interviewees. The 20 challenges that were nominated varied in terms of technologies, adversaries, and external factors.

Experts' evaluation of these challenges suggests that they generally fall into three clusters. A *top* cluster includes nine challenges that expert evaluators believed need urgent S&T investments, should be a major national defense priority, may have impacts with severe consequences, and overall were higher risk to the DoD and the United States. Experts tended to have more agreement on their ratings in this cluster. The *middle* cluster includes seven challenges that expert evaluators believed may have impacts with severe consequences but did not significantly need urgent S&T investment or to be a major defense national priority. A *bottom* cluster includes four challenges that expert evaluators did not believe had significance across the dimensions we measured.

While the 20 nominated challenges were diverse across types of technologies employed, impacts imagined, and focus of domains, a few broad patterns emerge in the overall nominations and three evaluation clusters. Challenges involving autonomous technologies, machine learning, or analysis of "big data" account for about half of the 20 challenges nominated. This may suggest that our nominating experts see these technologies as being an influential part of future conflicts with a peer adversary. The space domain also is included in about a third of the nominated challenges, potentially implying some importance in future domains. The patterns identified here could be influenced by the military background or focus of the expert nominators and evaluators. As discussed later in the chapter, our experts may place higher priority in areas of current military strategy or research.

No immediate pattern is clear for nominations or evaluations of challenges in terms of impact—that is, kinetic vs. nonkinetic or direct vs. indirect (e.g., attacks on U.S. forces vs. political instability leading to weakened U.S. forces). The lack of a pattern may suggest that our experts saw these types of impacts as equally important or, alternatively, that there is great uncertainty about future challenges.

Indeed, only three of the 20 challenges were rated as significant on our likelihood dimensions, with many mean ratings falling below the scale midpoint. A number of possibilities could explain this result. First, experts participating in the challenge nomination exercise may have chosen to nominate far-reaching challenges at the expense of plausibility. Indeed, we asked expert nominators to nominate challenges that were "not immediately intuitive . . . expressed in imaginative ways." A second theory is that, when deciding which challenges to nominate, expert nominators weighed the impacts more heavily than likelihood in their decisionmaking. The estimation of low-probability, high-consequence events as having greater risk than high-probability, low-consequence events is a common cognitive bias in risk perception (e.g., Slovic et al., 1986). Another possibility is that the challenges, as presented, were too specific to be estimated as highly likely. Relaxing a couple of the challenges' assumptions, for example, could impact their Likelihood ratings. Finally, it is also possible that the two criteria we used in our composite measure were not fully congruent with a likelihood estimation. Indeed, we heard during our formative interviews that attempting to estimate the likelihood of these challenges would result in unreliable results, given the deep amount of uncertainty inherent in each. The two criteria discussed barriers to a challenge being realized. Admittedly, a reliable estimation of the likelihood of any event requires knowledge of more than just the barriers to its occurrence.

For example, an adversary with great means may still decide to exploit a technological U.S. weakness even if it is difficult, given the impact is severe enough. Overall, our two barrier-related criteria may inform a likelihood estimation but alone are likely not sufficient. Instead, this composite may be better assessed as an *indicator* of the likelihood of one of these challenges being realized.

Given the clustering of challenges and their statistical significance against our five final criteria, a few different interpretations are possible when deciding which should receive priority in terms of being radically altering. One interpretation is to prioritize the nine challenges in the top cluster. Expert evaluators saw these as having an urgent need for DoD S&T resources and as potential national defense priorities. However, it is possible these evaluations are biased by current DoD efforts. Certainly, our expert evaluators either perform research for DoD or (in the case of military fellows placed at RAND) are members of one of the services. If DoD is currently considering challenges such as these, our expert evaluators are likely well informed about aspects of them. Their better-informed position could explain why we see experts in more agreement of their evaluation of the top cluster and potentially cause them to weigh these challenges more heavily in their assessments. If this theory holds, then another interpretation of our results would suggest that the seven challenges in the middle cluster should receive priority. Possibly they are not as well known to our expert evaluators, as suggested by larger dispersions around expert mean ratings. Therefore, experts may not have seen these challenges as important in terms of urgent S&T investments or national defense priority. Still, each of these challenges was rated as having significant impact on the DoD or the United States if realized. Therefore, priority might be given to these middle challenges in order to better understand and estimate their significance.

While the remaining four challenges were not shown to have significant ratings across our criteria, their presence in the original nominations does suggest at least some consideration by OUSD(R&E). Possibly priority need not be given to them at this time. However, the DoD may benefit from monitoring technological advances and adversary activity in these areas. For the top and middle clusters, though, deciding which to prioritize likely depends on the intentions of the OUSD(R&E) sponsor. If the intent is to explore new, less intuitive challenges, the middle cluster should be prioritized. If instead, the intent is to deepen knowledge in challenges potentially already being considered by DoD, the top cluster should be prioritized.

We note, however, that some caution should be applied in explicit consideration of the results and interpretation of challenge results presented in this report. These results are intended to inform findings and recommendations discussed in the main report of the project. The credibility, accuracy, and scope of these challenges will be further scrutinized for that purpose. In their current form, the challenges may be limited by the scope of the structured elicitation exercise. While the original nominated challenges were culled and slightly modified for inclusion in the evaluation survey, their scope, assumptions, and significance were deliberately unaltered by the project team. The credibility and importance was left to be interpreted by the evaluating experts. We refer the reader to the main report for further discussion.

Generalized Use of the Mixed-Method Approach for Targeted "Crowdsourcing"

We developed a three-stage expert elicitation approach as a way to crowdsource the most radically altering challenges in a targeted manner. Application of the generalized approach may be beneficial for other idea crowdsourcing tasks. The approach has two clear strengths compared with other approaches used for eliciting technology-initiated challenges. First, the initial use of exploratory qualitative interviews to develop the specific materials and methods to nominate and evaluate ideas improves the reliability and credibility of final results. Second, implementing a well-structured approach enhances transparency and allows for researchers to explore whether experiment design changes can have an impact on outcomes.

However, a number of limitations of the approach are worth noting. First, our entire decision framing for the nomination and evaluation forms is based on interviews with a small number of experts.

The sensitivity of results to the experts chosen for the sample should always be a consideration in any expert elicitation conducted. Indeed, we make no claims that our results are representative of all experts with similar domain expertise. Our experts were chosen exclusively from RAND's population, as prescribed by the 2040 Challenges project objective. Results from any of the three phases of the elicitation could be different if experts from outside of RAND participated. Within RAND, we saw very few significant differences in challenge evaluations between RAND experts and military fellows. However, the actual challenges that were nominated for evaluation and the evaluation criteria themselves could have been different if other experts (from within RAND or outside) had been chosen. Nevertheless, we did find great commonalities in themes extracted from the interviews[1] and the 40 initial challenges nominated (which because of these commonalities were able to be reduced by half). Overall, any crowdsourcing approach emulated from the one presented here must, importantly, consider the expert sample chosen.

Another limitation to our approach is a common one across all types of surveys—survey length. As survey length increases, respondent participation can decrease (either by deterring respondents from accepting an invitation or by causing them to quit before fully completing the survey) and cognitive survey fatigue can set it (Tourangeau, 1984). Both can reduce the quality of survey results. To combat these issues, we attempted to limit our challenge evaluation form to a length that could be completed in two hours or less. This required us to reduce the number of challenges evaluated, the length of challenge descriptions, and the number of evaluation statements. Among those three limitations, we found the number of evaluation criteria the most difficult to limit. We acknowledge that certain dimensions that might make a challenge more or less important to prioritize may have been omitted or combined with others to ensure an appropriate survey length. If our approach is adopted for other targeted exercises for crowdsourcing ideas, these considerations will surely emerge for the evaluation task.

---

[1]   However, our qualitative analysis of these interviews used a streamlined approach, which may have reduced our ability to objectively identify and analyze these themes to some extent.

Exploratory Interview Materials

Homework Assignment for Expert Interviews

As a reminder, the purpose of this project is to illuminate technological, operational, organizational, and cultural challenges that have the potential to radically alter the strategic and operational environments the U.S. military may encounter by 2040. We are specifically interested in challenges that might emerge because of the confluence of multiple technological developments. While these technologies can still be emerging, they should at least currently be at a stage in which they have been validated in a laboratory environment. The challenges can take the form of organizational and cultural as well as technical developments but should have operational or battlespace consequences/implications. We will be presenting our final list to the project sponsor.

As you answer the questions below, please jot down notes about your answer.

1. Choose one challenge that you think might or should be on our final list that we present to the sponsor. Please think through:
    a. What is the challenge?
    b. What sequence(s) of events could lead to that challenge? How likely is this (are these) sequence(s) of events?
    c. Why does that challenge have the potential to radically alter the strategic and operational environments the U.S. military may encounter by 2040? What are the potential impacts of the challenge and how likely are those impacts?

2. Choose a second challenge that you think might or should be on our final list that we present to the sponsor. For this challenge, please pick one that you think is quite different from the first one you chose.
   a. What is the challenge?
   b. What sequence(s) of events could lead to that challenge? How likely is this (are these) sequence(s) of events?
   c. Why does that challenge have the potential to radically alter the strategic and operational environments the U.S. military may encounter by 2040? What are the potential impacts of the challenge and how likely are those impacts?
3. Compare the two challenges you have brainstormed.
   a. Which challenge has the greatest potential to radically alter the strategic and operational environments the U.S. military may encounter by 2040? Why?
   b. What characteristics or factors make the one you chose more "radically altering" than the other?
4. Now think about a challenge that has already occurred in the recent past that you consider to have been "radically altering."
   a. What was the challenge?
   b. What sequence(s) of events led to that challenge?
   c. How did this challenge radically alter the strategic and operational environments of the U.S. military? What were the impacts of the challenge?

Expert Interview Protocol

Introduction (10 min)

Thank you for taking the time to speak with us today. Before we begin this interview, I'd like to briefly tell you a bit about the project and how this interview will be used. Did you receive the email from me that provides a brief description of the project goal? Can you open that up? We'd like you to use this for reference as we go through this discussion.

This project is sponsored by the Office of the Under Secretary of Defense for Research and Engineering (OUSD[R&E]) and seeks to illuminate technological, operational, organizational, and cultural challenges that have the potential to radically alter the strategic and operational environments the U.S. military may encounter by 2040. We are specifically interested in challenges that might emerge because of the confluence of multiple technological developments. That is, how could a sequence of technology developments lead to major challenges to the U.S. military in 2040? We are referring to these types of challenge as "game-changing challenges." You can always refer to the project goal description I sent ahead of time if you need to refamiliarize yourself with what we mean.

To develop the game-changing challenges we will present to the sponsor, we are enlisting the help of a variety of internal RAND experts. This interview is the first stage of the process. We are talking to a small subset of experts to learn two primary things:

- How do experts think about game-changing challenges? What problem framing and language do they use?
- What makes a challenge "game-changing"? For example, if we state that one challenge is more game-changing than another, what factors are being considered?

Once we have a better understanding of these foundational concepts, later stages of the project will ask experts to (1) nominate and describe what they believe are the top challenges and (2) provide inputs to help us rank those candidate challenges in terms of how game-changing they could be. So in this stage, we are really just trying to wrap our heads around the problem, and we have asked you to speak with us today to help inform that.

Soon, I'll begin asking you some open-ended questions. Please try to answer them as best you can. There are no right or wrong answers. However, I may at times ask follow-up questions or move us on to another topic. We only have a short amount of time, so I need to make sure we stay on task and get the most useful information. We value your perspective—that is why you are here today!

This interview will last approximately 1.5 hours. I will be asking you questions, but my colleagues will also be listening in and may interject at times. They will be taking notes about our conversation today. But it would also be very helpful to audio-record the interview so we can refer to it later if need be. No one outside of the project team will have access to the notes or audio-recording. We may use your ideas or direct quotes from this interview (for example, in later stages of the process or in our final briefing or report), but they will never be attributed to you and will not be identifiable. You can stop this interview at any time, with no repercussions, and you will receive a [time-charging code] for the 1.5 hours.

Before we begin, do you have any questions? Is it OK with you for us to audio-record this interview?

Great, then let's get started.

Challenge-Focus Questions (25 min)

So let me reiterate the purpose of this project: to illuminate technological, operational, organizational, and cultural challenges that have the potential to radically alter the strategic and operational environments the U.S. military may encounter by 2040. We are specifically interested in challenges that might emerge because of the confluence of multiple technological developments.

To get your mind calibrated, let's start with some more specific questions. Off the top of your head, can you tell me a bit about one game-changing challenge that you think might or should be on our final list that we present to the sponsor?

- How did you come up with this? What was going through your head as you were deciding what to talk about?
  - [If they don't mention the following add these probes.]
  - In the example, what specifically is the "challenge"?
  - Can you describe the sequence of events that could result in this challenge? Are there other sequences of events that could lead to a similar result?
    - Can you talk a bit about how likely you think this/these scenarios are and why you think that?
  - Why is that challenge game-changing?
    - What factors did you consider when coming to that conclusion?

- Were there certain assumptions you made when deciding that it was game-changing? What were they?
  - In your decision to choose this challenge, how did you account for the likelihood related to the sequence of events leading up to this challenge?
  - How did this likelihood weigh in against the other factors you considered when choosing this challenge?
- What was hard about coming up with this example?
  - What information do you believe would have been helpful to have upfront to best answer the question?

Problem Framing (20 min)
Okay, now let's take a step back and talk more generally.

Can you tell me a little about your area of expertise?

How do you believe your area of expertise relates to the goals of the project?

- When we are talking about a game-changing challenge in your area of expertise, what does that mean to you? How would you describe what is meant by a "game-changing challenge"?

Now, stepping outside of your specific area of expertise, do you think the term "game-changing challenge" takes on a different or maybe broader meaning? If so, how?

Now, I'd like to ask you a few questions that should directly help us in our next stage of the project—when we reach out to a larger set of RAND experts to nominate and describe the top challenges we should consider in the project.

- How do you think we should ask experts for this information? What language should we use? What information should we provide to them?
- The word challenge is quite vague. Thinking back to the example challenge you gave earlier, is there a correct way to scope a chal-

lenge? Can a challenge be too specific or too broad? How do we ask for adequately scoped challenges from our experts?

- For this challenge generation exercise, we want to obtain a list of challenges that is broad enough to encompass a diverse set of challenges but not so broad that it becomes irrelevant or out of scope. Given that:
  - What areas of expertise are most important to include when identifying our experts?
  - How should we choose experts to get diverse viewpoints? For instance, are there qualities past the area of expertise that we should be thinking about?
  - Should we specify a number of challenges that experts should generate (e.g., top five) or ask for their nominations in some other way?
  - We want challenges that are the result of an uncertain sequence of technological developments. Do you think we should focus on eliciting the most consequential challenges, regardless of how likely the sequence leading up to them is? Should we explicitly try to elicit some less consequential challenges that are more likely to occur? How do we ensure we are eliciting challenges over the wide range of likelihoods? For instance, ones that might be underrated or ignored or possibly be a minority viewpoint?
  - Is there anything else you think we should know as we develop a call to experts for them to nominate challenges?

Evaluating/Comparing Challenges (20 min)

Now let's switch gears a little bit and talk more about what makes a challenge game-changing or one challenge more game-changing than another. Earlier in this interview you discussed a particular challenge. Now, I would like you to discuss another game-changing challenge that you think might or should be on our final list that we present to the sponsor. But I'd like you to pick a challenge that you think is pretty different from the first one you chose. Can you describe that challenge to us, the sequence of events leading up to it, and what makes it game-changing?

We have discussed two very different challenges today and now I would like to discuss how to compare the two.

- Is one challenge overall more game-changing than the other? If so, why?
- Now focus on the original challenge you discussed. Are there certain characteristics or qualities of this challenge that you think make it more game-changing than the other? What are those characteristics? Please describe your thinking about them.
- Now, flip the question. Focus on your second challenge. Are there certain characteristics or qualities of this challenge that you think make it more game-changing than the other? What are those characteristics? Please describe your thinking about them.
- When determining which challenge is more game-changing:
  - Were there certain assumptions that you needed to make? What were they?
  - How did you account for the likelihood of the events leading up to these challenges? How did that factor in to your decision-making about which challenge was more game-changing?

Let's talk a little more broadly about this idea of comparing challenges. As a reminder, a later stage of our research is going to ask a larger set of experts to evaluate a candidate list of challenges in some way to determine a rank ordering in terms of how game-changing they are.

- How would you recommend we ask experts to evaluate or compare these challenges?
- Are there certain qualities or characteristics we should ensure they consider when evaluating how "game-changing" a challenge is?
- Do some of these characteristics have thresholds that are worth considering? For example, if a challenge is so high on this one characteristic, it really doesn't matter too much how it fares on the others such that you would consider it a top game-changer. Please explain.
- We expect that each challenge being considered will have different likelihoods associated with it and that our candidate list

will partially depend on future unknown scenarios. Do you have thoughts on how to account for the likelihood of the events leading up to the challenge in the overall game-changing "equation"?

- Is there anything else you think we should know as we develop a call to experts for them to nominate challenges?

Historical Calibration (10 min)

One final set of questions now, in which I would like to switch gears one more time. This time I would like you to describe a challenge that has already occurred in the recent past that you consider to have been game-changing.

- Tell me the story behind how that challenge emerged (e.g., the sequence of events) and what the consequences were to it being realized.
- Why do you consider it to have been game-changing?
- Now think back to some of our earlier discussions about how to elicit ideas about challenges and evaluate how game-changing they are. Based on this historical example:
  - What would have been the proper way to elicit this type of challenge from experts? How would you have asked them to generate this idea?
  - What factors would have helped to evaluate how game-changing this challenge really was?

Final Thoughts (5 min)

We are almost done. Now that we have run through this interview, do you have any other thoughts or comments that you would like to share with us? Is there anyone else that you believe would be helpful for us to interview at this stage in the process or involve in later stages of our project?

Thank you very much for your time.

Challenge Nomination and Evaluation Materials

Challenge Nomination Form

Please use this document to nominate and describe two "radically altering" 2040 challenges that you believe the RAND team should present to the OUSD(R&E) sponsor. In a future stage of the process, we will ask a set of experts to prioritize all the nominated challenges to determine which should be elaborated and presented to the sponsor. The following instructions provide specific guidelines for the types of challenges that meet the project scope and the needed information. Please read them carefully.

Instructions
What types of challenges are we looking for?
The challenges you nominate *must* have the following characteristics:

- Be at least partially initiated by two or more technological developments that come together in some way and that (to be credible causes of concern in 2040) have at least been proven in a laboratory environment as of today.
- Have the potential to radically alter the strategic and/or operational environment the U.S. military may encounter by 2040.
- Generate impact that can be technological, operational, organizational, and/or cultural.
- Be capable of inspiring/motivating a specific story or narrative that has the potential to occur by 2040. The challenge should be able to be expanded (at a later date) to provide enough detail and/or specificity to allow the sponsor to perform tabletop exercises.

- Be not immediately intuitive: the sponsor is looking for ideas he hasn't heard before (he is immersed in data and analyses about technological trends), expressed in imaginative ways that can get people with operational responsibilities thinking and planning in new ways.

The challenge you nominate *may* have the following characteristics:

- Inspire/motivate additional circumstances or scenarios that in concert with the technology-based developments lead to the challenge.
- Involve technologies that may be:
  - Currently used or intended for use in commercial, civilian, and/or military situations.
  - Potentially adopted by the United States, its allies, and/or its adversaries.
- Occur through intentional or unintentional use of the technologies, or through the lack of use of those technologies.

What do you need to do?
We are asking you to nominate two challenges. One challenge should be in your area of expertise (e.g., you have done related research). We ask that the second challenge be one that is outside your area of expertise, but one for which you feel knowledgeable enough to provide the information being elicited. Both challenges should be ones that you believe (1) have the potential to be "radically altering" and (2) should be presented to the project sponsor.

What information do you need to provide?
The questionnaire on the next page asks you to provide the following information for each challenge:

- A short title for the challenge
- High-level summary of the challenge
- A list and description of the technologies initiating the challenge and how they interact

- Description of the external factors or assumptions involved in the challenge
- Description of the impact of the challenge
- Why you believe the challenge is important to present to the sponsor

Please only nominate challenges for which you can provide this information. Answers don't need to be long, and we ask for each challenge to take less than one page to describe.

Challenge #1 (within your area of expertise)
Provide a short title for your challenge:

Provide a high-level summary of the challenge in no more than a short paragraph.

List and briefly describe the technologies that could initiate this challenge and how they interact.

Briefly explain external factors or assumptions involved in the challenge.

Briefly describe the impact of the challenge.

Why is this challenge important to present to the sponsor?

Challenge #2 (outside your area of expertise)
Provide a short title for your challenge:

Provide a high-level summary of the challenge in no more than a short paragraph.

List and briefly describe the technologies that could initiate this challenge and how they interact.

Briefly explain external factors or assumptions involved in the challenge.

Briefly describe the impact of the challenge.

Why is this challenge important to present to the sponsor?

Challenge Evaluation Form (Excerpt)

On the following pages, you will be presented with a number of challenges that have the potential to radically alter the strategic and/or operational environments the U.S. military may encounter by 2040. These challenges were nominated by your RAND colleagues to be presented to the OUSD(R&E) sponsor. Your job is to help us evaluate this long list of challenges to determine which should be elaborated on and presented to the sponsor. All the challenges below meet the following criteria, as scoped by the sponsor:

- Are at least partially initiated by two or more technological developments that come together in some way and that (to be credible causes of concern in 2040) have at least been proven in a laboratory environment as of today.
- Have the potential to radically alter the strategic and/or operational environment the U.S. military may encounter by 2040.
- Generate impact that can be technological, operational, strategic, organizational, and/or cultural.

Please read each challenge carefully. Each challenge description will contain five parts: a summary, a description of the technologies involved, any external factors or assumptions involved, the impact and the importance to OUSD(R&E). After reading the challenge, you will be presented with following seven statements:

- DoD is *unlikely* to detect this challenge prior to it being realized *or* to prevent it from being realized even if it is detected (e.g., because of a lack of ability or willingness, or other barriers).
- *Other than* barriers to DoD detection and prevention, there are *very few* barriers that could keep this challenge from being realized (barriers may include, for example, lack of technology readiness or adversary ability/intent to adopt technology, forces that change external factors, U.S. institutional or cultural incentives).
- If this challenge is realized, it will be difficult for DoD to mitigate its impact (e.g., through authorities, capability, knowledge, costs, effort, or time).

- *Left unmitigated*, this challenge will severely limit (directly or indirectly) DoD's options for performing operations.
- *Left unmitigated*, this challenge will severely negatively impact (directly or indirectly) U.S. national security (e.g., via changes to economy, culture, politics, military operations).
- If DoD does not invest soon, *specifically in appropriate S&T areas*, the opportunity to reduce the risks of this challenge being realized will have passed.
- Preparing for this challenge should be a major national defense priority.

After reading each statement, you will be asked to indicate your level of agreement with each, on a scale from "strongly disagree" (= 1) to "strongly agree" (= 7). When providing your answer, please consider the challenge in its entirety. That is, for the purposes of this evaluation, a "challenge" is defined as the holistic narrative presented (i.e., technologies, assumptions, external factors, and impact).

We expect there to be some judgment in this task, but if you truly are unable to provide an evaluation of a statement, you can also answer "don't know." As each challenge is unique, a "not applicable" option is also available if you believe the statement cannot be related to the challenge. We will use your ratings of the statements to determine which challenges to present to the sponsor. If you have any comments about a specific challenge that you would like to provide, space is provided on the last page of this form.

After reading the challenge, you will also be asked to indicate your level of expertise with that specific challenge on a scale from "no knowledge" (= 1) to "nationally recognized expert" (= 5).

At the very end of the document, you will be asked to indicate your level of agreement with two final statements about the evaluation criteria used in the form. Please answer on a scale from "strongly disagree" (= 1) to "strongly agree" (= 7).

Thank you for your time and help! The following pages contain 20 challenges. We recommend splitting up your evaluation of these challenges over the course of a couple days to ensure you do not become cognitively fatigued.

Before getting started, please answer some basic questions about yourself.

1. Name _____
2. In which RAND Division (e.g., NSRD, PAF, Arroyo, HSOAC, SEP) have you worked the most over the last three years? (Please provide only one division; your best guess is fine.) _____
3. For the following areas of expertise, please also indicate your level of expertise on a scale from "no knowledge" (= 1) to "nationally recognized expert" (= 5).

	Area of Expertise	No knowledge	Layperson knowledge	Some professional experience	Lots of professional experience	Nationally recognized expert
Military domain	Cyber	❑	❑	❑	❑	❑
	Air	❑	❑	❑	❑	❑
	Space	❑	❑	❑	❑	❑
	Land	❑	❑	❑	❑	❑
	Sea	❑	❑	❑	❑	❑
	Special operations	❑	❑	❑	❑	❑
Region/ adversary	Example 1	❑	❑	❑	❑	❑
	Example 2	❑	❑	❑	❑	❑
	Other (specify ___)	❑	❑	❑	❑	❑

End of Anonymity

Summary: By 2040, it will be increasingly difficult for individuals to operate anonymously and to assume fictitious identities. For example, low-cost retinal and fingerprint scanners connected to comprehensive databases, as well as facial recognition technology, might make it challenging for someone to assume a new identity. Rapid, low-cost genome sequencing and genotyping, accompanied by growing public databases of DNA sequences, could further complicate the problem by revealing a person's relatives and hence her or his backstory. China is already demonstrating how a state can control social interactions and political expression through the control of Internet applications and use of biometrics and online surveillance tools. These technological capabilities clearly offer military and political advantages to China and other states that choose to adopt them.

Technologies Involved: Two main sets of technologies interact to create this challenge. The first set includes technologies that provide data about a person: scanners and sensors (e.g., retinal, fingerprint, facial) and genotyping and genomic sequencing technologies. The second set includes technologies allowing rapid analysis of data created by the first set (e.g., machine learning, high-speed computing, and databases). Both sets of technologies are becoming faster and less expensive to use.

External Factors or Assumptions Involved: The largest assumption is that the Chinese population will continue to acquiesce to the domestic application of these technologies, allowing the Chinese government to continue to refine and train their AI algorithms for applications focused on identifying and targeting individuals and groups.

Impact: First, the use of these technologies by a state would substantially expand the tools that state could use to influence geopolitical positioning. It would also provide the state deploying the technologies with types of power over other states' populations that might be difficult to counter through military means. Second, these technologies could render the collection of human intelligence next to impossible by eliminating the possibility of operating anonymously.

Importance to USD(R&E): Because of the pervasiveness and extra-territorial nature of these technologies, this challenge could undermine the successful execution of U.S. diplomatic strategies, alliances, and military concepts of operation. The behavior of populations and states under threat by our adversaries could be radically different when that threat includes ubiquitous surveillance.

Evaluation Statements	Strongly Disagree 1	Disagree 2	Somewhat Disagree 3	Neither Agree nor Disagree 4	Somewhat Agree 5	Agree 6	Strongly Agree 7	N/A	Don't Know
1	DoD is *unlikely* to detect this challenge prior to it being realized or to prevent it from being realized even if it is detected (e.g., because of a lack of ability or willingness, or other barriers)	☐	☐	☐	☐	☐	☐	☐	☐	☐
2	*Other than* barriers to DoD detection and prevention, there are very few barriers that could keep this challenge from being realized (barriers may include, e.g., lack of technology readiness or adversary ability/intent to adopt technology, forces that change external factors, U.S. institutional or cultural incentives)	☐	☐	☐	☐	☐	☐	☐	☐	☐
3	If this challenge is realized, it will be difficult for DoD to mitigate its impact (e.g., through authorities, capability, knowledge, costs, effort or time)	☐	☐	☐	☐	☐	☐	☐	☐	☐
4	*Left unmitigated,* this challenge will severely limit (directly or indirectly) DoD's options for performing operations	☐	☐	☐	☐	☐	☐	☐	☐	☐
5	*Left unmitigated,* this challenge will severely negatively impact (directly or indirectly) U.S. national security (e.g., via changes to economy, culture, politics, military operations)	☐	☐	☐	☐	☐	☐	☐	☐	☐
6	If DoD does not invest soon, *specifically in appropriate S&T areas,* the opportunity to reduce the risks of this challenge being realized will have passed	☐	☐	☐	☐	☐	☐	☐	☐	☐
7	Preparing for this challenge should be a major national defense priority	☐	☐	☐	☐	☐	☐	☐	☐	☐

Expertise Self-Evaluation	No knowledge 1	Layperson knowledge 2	Some professional experience 3	Lots of professional experience 4	Nationally recognized expert 5
Indicate your level of expertise for the technologies in this challenge	☐	☐	☐	☐	☐
Indicate your level of expertise for other aspects of this challenge (e.g., external factors, adversary, its impact)	☐	☐	☐	☐	☐

Selected Nominated Challenges

Geoengineering: Geoengineered Climate Warfare

Summary
Serious discussions on geoengineering methods for moderating greenhouse gas–based warming and its impact have been moving from the periphery of scientific discourse to center stage; however, the development of a global regime for governing and regulating the deployment of these methods has lagged far behind the rapid development of the associated technologies. The proposed challenge comes in the form of unilateral action by a determined adversary to either deploy one of these methods without proper oversight and safeguards or to sabotage, hijack, or otherwise manipulate an already-deployed strategy. Both might have potentially cataclysmic impacts.

Technologies Involved
Approaches include reflecting solar radiation (e.g., by putting mirrors into space or reflective particles into the atmosphere), sucking carbon dioxide out of the atmosphere, sequestering large volumes of carbon dioxide underground, implementing ocean fertilization, and conducting afforestation.

External Factors or Assumptions Involved
The first assumption is that the nations of the world will not create an enforceable global governance system for geoengineering. The second assumption is that these technologies will mature sufficiently over the next two decades to enable their economic deployment at scale.

Impact

Weaponizing any of these geoengineering technologies could have catastrophic consequences for the global climate system, potentially disrupting atmospheric circulation patterns and the distribution of ocean and land temperatures. It may also cause changes in precipitation and storm generation. These changes could spur mass migration and disrupt agricultural production more acutely and on a larger scale than the incremental climate change effects earlier in this century.

Importance to OUSD(R&E)

Such changes could create new regions of political instability. This challenge could also have a profound impact on the efficacy of military strategies founded on assumptions about a predictable operating environment. It could also degrade the functionality of some military platforms and sensors and disrupt deployment of capabilities in certain parts of the world.

Minefields: Improved UAS, Sensors, Energy Storage, and Communications Enable Durable and Effective Minefields

Summary

In this scenario, adversaries seeking to attrit and limit the movements of U.S. forces can deploy more capable minefields to do so. This capability is enabled by other emerging capabilities that allow networking of small, inexpensive sensors and short-range projectile weapons; these networks can collectively achieve higher levels of lethality and wider area effects than traditional mines. Such minefields are also particularly challenging to breach or degrade. Weapons and sensors can be deployed on low-visibility fixed nodes or unmanned vehicles at relatively low cost. Moreover, it is possible that mining may be extended to the air domain. Distributed sensors placed on small lighter-than-air vehicles could linger in the environment alongside small swarms of inexpensive, autonomous UASs poised to crash into or otherwise target warplanes.

Technologies Involved

Improved autonomous unmanned systems (e.g., better cost, speed, payload, and endurance); sensors and algorithms for comprehending sensor outputs; energy storage, efficiency, and harvesting; communication and networking protocols and technologies; and related materials (e.g., through additive manufacturing).

External Factors or Assumptions Involved

U.S. adversaries have a keen interest in creating "no-go" zones that U.S. forces are unable to transit or operate in without incurring high losses. Inexpensive, automated, distributed minefields could be an effective means of thwarting U.S. power projection.

Impact

More durable and effective minefields could attrit substantial numbers of U.S. ground vehicles, surface ships, submarines, and possibly even warplanes. In addition, they could delay or disrupt operations by preventing U.S. forces from accessing critical geographies.

Importance to OUSD(R&E)

Minefields are often dismissed as a threat because they appear less menacing and technologically sophisticated than, for example, hypersonic missiles. However, adversaries can use new technologies to exacerbate existing, less sophisticated threats. DoD runs the risk of being unprepared to counter an upgraded minefield threat if it overlooks this possibility.

Autonomous Unmanned Aerial Systems: The Proliferation of Fully Autonomous UAS and UAS Swarms

Summary

In this challenge, lethal, autonomous UAS and UAS swarms have become widely available to all near-peer competitors, rogue states, and well-funded nonstate actors. Proliferated UAS and UAS swarms may be able to overwhelm existing defenses, execute kinetic attacks, conduct cyberattacks, and conduct intelligence, surveillance, and reconnaissance

(ISR) missions. Proliferated autonomous systems may also degrade escalation control by removing steps involving human discretion.

Technologies Involved

Battery technology including Li-metal, flow, and solid-state batteries; sensors; algorithms; communication protocols (e.g., 5G/6G, Disruption Tolerant Mesh Networks); navigation approaches such as inertial navigation, machine vision, and simultaneous localization and mapping; and sophisticated stabilization systems, feature detection, and obstacle avoidance capabilities.

External Factors or Assumptions Involved

It is assumed that autonomous swarm algorithms will be fully realized by 2040. AI will allow target recognition/discrimination and kill decisions, secure interfaces for swarm coordination will be difficult to jam or dazzle, and low-latency communications at 5G speeds will be either deployed or rapidly deployable to allow computing power to be off-loaded from an individual small UAS to nearby (edge) processors.

Impact

Lethal autonomous swarms may require large changes in U.S. force protection posture. For example, fixed base locations provide an easy target for UAS-enabled kinetic attacks. In forward-operating areas, lethal autonomous swarms may be able to recognize and attack U.S. and allied personnel. These technologies will also present a new and difficult challenge for dispersed ground forces, which will be at a speed-of-response disadvantage relative to lethal autonomous systems.

Importance to OUSD(R&E)

The evolution of UAS and UAS swarms, including the use of small systems, in the commercial sector is accelerating the development of UAS-based weapons platforms beyond what traditional military-only development could accomplish. By 2040, small, lethal, autonomous UAS will be widely available; swarms of such systems will effectively become an entirely new type of weapon with the potential to change wars.

Virtual Reality: The Impact of Virtual Reality on Training in the Armed Services

Summary

The popularity and increased technical sophistication of video games have changed how young people in the United States spend their leisure time. As virtual (and augmented) reality technologies become more affordable and widespread, we may face another shift in how young people spend their free time. The proliferation of virtual reality technologies has the potential to fundamentally change how the U.S. military trains personnel and may significantly affect the fitness and skill sets of the population from which the DoD recruits.

Technologies Involved

Video game, virtual reality, and augmented reality technologies including consoles, heads-up displays, headsets, gaming software, sensors, graphics, and peripheral devices.

External Factors or Assumptions Involved

This challenge assumes that video game and virtual reality technologies will continue to expand in popularity and technical capability, leading to an increase in the number of leisure activities done "virtually" rather than conventionally. It also assumes that practical applications of virtual and augmented reality will proliferate (affecting social, economic, and military activity).

Impact

The DoD could be crippled if its training practices cannot account for population-level changes in the skill set and experience of potential recruits. For example, the U.S. Air Force aircraft maintenance community is already facing challenges in training new recruits who, unlike in previous generations, enter service with no hands-on experience in any sort of trade or repair work. The inability to effectively populate and train the U.S. armed forces could have far-ranging consequences.

Importance to OUSD(R&E)

Video games and virtual reality are already changing how much of the U.S. youth population spends its leisure time, at the expense of hobbies and skills that have historically had value to the armed forces. As these technologies become both more immersive and more pervasive, the U.S. armed forces may soon face a population that does not have needed skills and that is accustomed to learning new skills in a way (e.g., technology-aided) that lies outside the traditional armed forces training paradigm. Although the Defense Advanced Research Projects Agency (DARPA) has a limited history of applying information technology to training, a much broader shift in how the services train may be needed to avoid shortages or deficiencies in critical skills.

Spoofing Vulnerability: Reliance on Automated Systems Creates Vulnerability to Spoofing and Data Manipulation

Summary

As the national security enterprise—and society as a whole—increasingly relies on automated systems for decisionmaking support, associated decisions may be exposed to potential interference. Development of these automated decisionmaking methods may be aided by an increased understanding of the physiology behind decisionmaking built through advances in neuroscience. These concurrent technological and scientific developments could create opportunities for an adversary to manipulate data in a more targeted way, shaping national security decisionmaking and degrading the U.S. military's ability to operate.

Technologies Involved

Machine learning algorithms and advances in neuroscience and brain imaging.

External Factors or Assumptions Involved

This scenario assumes that the use of automation in decisionmaking will grow. It also assumes that the ability to compromise this automation, through cyber or other types of disinformation campaigns, will continue to grow and democratize.

Impact

If the U.S. military increasingly uses automated decisionmaking processes for its operations, it will leave itself vulnerable to manipulation.

At best, this will decrease trust in automated decision-support tools as well as reduce their efficacy. At worst, this will eventually facilitate virtually undetectable manipulation by our adversaries.

Importance to OUSD(R&E)

OUSD(R&E) is likely already considering challenges related to automation and trust in automated tools; however, it may not be considering the novel ways to manipulate data and algorithms that will be developed concurrently with these tools. As both automation and manipulation technologies develop, there may be a sort of arms race between the development of manipulation techniques and the development of machine learning algorithms to detect this manipulation. The United States may lose this race if the efficacy of manipulation techniques is bolstered by advances in neuroscience—an arena where our adversaries are investing more (for malign purposes).

Quantum Computing: Quantum Computing Creates a World Without Encryption

Summary

Quantum computing is currently emerging from laboratories to practice. While many discussions around quantum computing focus on its transformative benefits, its likely transformative effect on encryption is more concerning. The security challenges that quantum computing will create are already drawing attention, and we will likely start to feel the real impacts of those challenges shortly before 2040. Those with access to quantum computing may be capable of breaking encryptions—a necessary tool in defending computer networks—at scale. As access to quantum computing increases, current encryption protocols could be broken en masse by an adversary.

Technologies Involved

Public-key encryption relies on mathematical relationships to protect messages and data. While these relationships can all be broken in theory, the required computation time typically makes this infeasible in practice. Quantum computing—the nontraditional use of quantum mechanics for computation—may increase computing power sufficiently to make the rapid breaking of many encryption techniques feasible for an adversary.

External Factors or Assumptions Involved

This challenge assumes that continued technological progress will make quantum computing available and accessible and that novel, quantum-resistant encryption solutions will not be developed prior to realization of the challenge.

Impact

By defeating encryption, quantum computing could enable widespread cyberattacks on U.S. military networks and information systems and prevent the United States from transmitting information securely. It would also facilitate widespread attacks on civilian (e.g., e-commerce) systems, and it would enable the unlocking of older classified information that has been protected by conventional encryption.

Importance to OUSD(R&E)

Holding an advantage in quantum computing and in quantum-resistant encryption would improve the overall U.S. security posture; holding a disadvantage would undermine that posture by increasing our vulnerability to cyber threats.

Human Adversaries: Overmatch from Augmented Human Adversaries

Summary

An adversary develops a cadre of superhumans who can analyze data, understand situations and possibilities, develop strategy and tactics, and act so quickly and with so much power and endurance that U.S. and allied assets cannot counter or defend against them. These superhumans can accomplish intelligence and elimination missions and can operate in engagements behind enemy lines to create uncertainty and direct forces. Developed in larger numbers and equipped with advanced weapons, they could be a deciding factor in conflicts.

Technologies Involved

High strength, elastic, lightweight materials for exoskeletons and prosthetics; brain-machine interfaces; improved human-computer interfaces; strength-, endurance-, and cognition-enhancing chemicals; implanted processors; and genetic engineering.

External Factors or Assumptions Involved

This scenario makes the following assumptions about the technologies that enable it: continued developments in exoskeletons and prosthetics (including brain control of these systems); significant advances in adversary understanding of neural chemistry, pharmacogenomics, genetic engineering, and related biotechnology; and active brain-machine interfaces. It will also be necessary for our adversaries to overcome likely societal aversion to this degree of human augmentation; however, genetic selection is already gaining acceptance, and countries with autocratic governments might be more willing to push the envelope.

Impact

The United States could find itself overmatched on the battlefield or restricted in terms of the military missions it could undertake.

Importance to OUSD(R&E)

Assessing the likelihood of this scenario requires tracking a diverse number of technological developments and continually evaluating the plausibility of their combination. It also requires an explicit focus on human-performance enhancement (by us and our adversaries).

Intelligent Robots: Adversaries Employ Intelligent and Merciless Robots

Summary

In this scenario intelligent, autonomous robotic weapons have proliferated. These robots have the capability to predict with reasonable accuracy what a human opponent might do and to counter that action with a set of actions that a human would not be able to accomplish. Furthermore, U.S. policy on the use of autonomous lethality still stresses the requirement for a human to intervene in the kill chain because of a variety of ethical and legal considerations. However, our adversaries are not so constrained and seek to leverage that asymmetry to their advantage. This asymmetry risks pitting U.S. troops against overwhelming quantities of "smart" machines that can act and react more quickly than humans.

Technologies Involved

Artificial Intelligence, reduced-SWaP (size, weight, and power) electronics (e.g., processing, sensors, communications), machine-to-machine collaboration, and mass manufacturing of complex technical systems.

External Factors or Assumptions Involved

This scenario assumes trends in AI and miniaturization continue and robotic warfare is embraced by our adversaries. It also assumes the U.S. military continues its policies with respect to fully autonomous kill chains.

Impact

The combination of intelligent autonomy and merciless behavior proposed in this scenario would bring about a new revolution in military affairs that could threaten U.S. military superiority across the spectrum of conflict. For example, in this scenario, the U.S military will operate much more slowly than our adversaries, giving those adversaries an operational advantage. In addition, if an adversary system were able to exploit a vulnerability in one of our systems during an engagement, we could lose multiple systems almost simultaneously before the vulnerability could be removed.

Importance to OUSD(R&E)

This challenge threatens U.S. military superiority across the full spectrum of conflict. To solve it, DoD must take an interdisciplinary approach that takes into account not just technology but also physiology, psychology, philosophy, ethics, law, and diplomacy.

Descriptive Statistics of Challenge Evaluation Results

This appendix presents three tables showing descriptive statistics of challenge ratings for the five criteria. Table D.1 provides mean and standard deviations, ordered by the mean ratings for the S&T criteria in descending order. To enable ease of comparison, Table D.2 preserves the ordering in Table D.1 but provides the rank ordering of standard deviations for each criterion, with the largest standard deviation receiving a rank of 1 and the smallest, a rank of 20. Table D.3 provides the minimum and maximum challenge ratings for the five criteria.

Table D.1
Mean and Standard Deviation of Challenge Evaluation Ratings

Challenge	S&T	Priority	Likelihood	Consequence	Risk
T1	5.30 ± 1.21	5.20 ± 1.13	4.11 ± 1.19	5.44 ± 0.88	4.90 ± 0.88
T2	5.30 ± 1.29	5.30 ± 1.31	4.02 ± 1.45	5.78 ± 1.13	5.09 ± 1.11
T3	5.13 ± 1.01	5.22 ± 1.13	4.17 ± 1.04	5.29 ± 1.03	4.86 ± 0.80
T4	5.00 ± 1.27	4.85 ± 1.33	4.44 ± 1.11	5.42 ± 1.10	4.98 ± 0.88
T5	4.93 ± 1.28	4.65 ± 1.33	3.90 ± 1.42	5.38 ± 1.14	4.77 ± 1.05
T6	4.85 ± 1.26	4.85 ± 1.42	3.49 ± 1.21	5.21 ± 1.06	4.52 ± 0.81
T7	4.81 ± 1.38	4.94 ± 1.32	3.65 ± 0.98	5.26 ± 0.97	4.60 ± 0.85
T8	4.73 ± 1.46	4.94 ± 1.41	4.53 ± 1.20	5.00 ± 1.22	4.81 ± 0.99
T9	4.71 ± 1.38	4.88 ± 1.47	3.66 ± 1.18	5.18 ± 1.13	4.59 ± 0.94
M1	4.48 ± 1.48	4.41 ± 1.69	2.97 ± 1.19	4.73 ± 1.50	4.07 ± 1.11
M2	4.41 ± 1.48	4.13 ± 1.48	3.48 ± 1.42	4.78 ± 1.28	4.22 ± 1.20
M3	4.33 ± 1.47	4.45 ± 1.44	3.73 ± 1.22	4.85 ± 1.36	4.39 ± 1.03
M4	4.32 ± 1.49	4.24 ± 1.52	3.77 ± 1.18	4.83 ± 1.32	4.45 ± 1.07
M5	4.26 ± 1.18	4.22 ± 1.48	3.47 ± 0.98	4.94 ± 1.26	4.30 ± 0.97
M6	4.13 ± 1.56	4.12 ± 1.29	4.14 ± 1.27	4.63 ± 1.22	4.41 ± 0.81
M7	3.94 ± 1.39	4.06 ± 1.24	4.41 ± 0.95	4.66 ± 1.16	4.51 ± 0.95
B1	3.76 ± 1.46	3.48 ± 1.62	3.84 ± 1.55	3.69 ± 1.27	3.73 ± 1.06
B3	3.59 ± 1.54	3.12 ± 1.52	2.62 ± 0.97	3.56 ± 1.55	3.18 ± 1.14
B2	3.34 ± 1.33	3.34 ± 1.60	2.85 ± 1.27	4.33 ± 1.40	3.79 ± 1.14
B4	3.32 ± 1.32	3.09 ± 1.27	3.14 ± 1.29	3.95 ± 1.32	3.63 ± 1.09

SOURCE: RAND analysis of expert evaluator survey responses.

Table D.2
**Ranking of Challenges by Evaluation Rating Standard Deviations
(in Descending Order)**

Challenge	S&T	Priority	Likelihood	Consequence	Risk
T1	18	19	11	20	15
T2	14	15	2	15	4
T3	20	20	16	18	20
T4	16	13	15	16	16
T5	15	12	4	13	9
T6	17	10	9	17	18
T7	11	14	18	19	17
T8	7	11	10	11	11
T9	10	8	13	14	14
M1	4	1	12	2	5
M2	5	6	3	7	1
M3	6	9	8	4	10
M4	3	4	14	6	7
M5	19	7	17	9	12
M6	1	16	7	10	19
M7	9	18	20	12	13
B1	8	2	1	8	8
B3	2	5	19	1	2
B2	12	3	6	3	3
B4	13	17	5	5	6

SOURCE: RAND analysis of expert evaluator survey responses.

NOTE: Lower ranking is associated with larger standard deviations. Larger standard deviations (lower rankings) are shaded red, median standard deviations are shaded yellow, and smallest standard deviations (higher rankings) are shaded green.

Table D.3
Minimum and Maximum Values of Challenge Evaluation Ratings

Challenge	S&T	Priority	Likelihood	Consequence	Risk
T1	[3, 7]	[2, 7]	[1.5, 6.5]	[3, 7]	[2.4, 6.6]
T2	[3, 7]	[2, 7]	[1.5, 7]	[2, 7]	[1.8, 7]
T3	[3, 7]	[3, 7]	[2, 6]	[3, 7]	[3.4, 6.2]
T4	[2, 7]	[2, 7]	[2, 6]	[3.3, 7]	[3.4, 6.6]
T5	[2, 7]	[2, 7]	[1, 6]	[3, 7]	[2.6, 6.4]
T6	[2, 7]	[2, 7]	[1.5, 5.5]	[3, 7]	[2.6, 6.2]
T7	[2, 7]	[1, 7]	[1.5, 6]	[3.3, 7]	[3, 6]
T8	[1, 7]	[1, 7]	[2, 7]	[2, 7]	[2.2, 6.2]
T9	[2, 7]	[2, 7]	[1, 5.5]	[3, 7]	[2.8, 6.2]
M1	[2, 7]	[1, 7]	[1, 5.5]	[1.7, 6.7]	[1.4, 5.8]
M2	[1, 7]	[1, 6]	[1, 6]	[1.7, 7]	[1.6, 6.6]
M3	[1, 7]	[1, 6]	[1.5, 6]	[2, 7]	[2.2, 6.6]
M4	[1, 6]	[1, 7]	[1, 6]	[2, 7]	[2, 6.6]
M5	[3, 6]	[2, 7]	[1.5, 5.5]	[2.7, 7]	[2.4, 6]
M6	[1, 7]	[2, 7]	[2, 7]	[2.3, 7]	[2.8, 5.8]
M7	[2, 6]	[2, 6]	[2, 6]	[2.7, 6.7]	[2.8, 6.2]
B1	[1, 6]	[1, 7]	[1.5, 6.5]	[1, 6]	[1.8, 6]
B3	[1, 6]	[1, 6]	[1, 4.5]	[1, 6.7]	[1, 5]
B2	[1, 6]	[1, 6]	[1, 5.5]	[1.3, 7]	[1.8, 5.8]
B4	[1, 6]	[1, 6]	[1, 6]	[1, 6.7]	[1, 6.2]

SOURCE: RAND analysis of expert evaluator survey responses.
NOTE: Values reported as [Minimum, Maximum].

References

Babbie, Earl R., *The Practice of Social Research*, Nelson Education, 2015.

Bernard, H. Russell, Amber Wutich, and Gery W. Ryan, *Analyzing Qualitative Data: Systematic Approaches*, Sage Publications, 2017.

Bidwell, Christopher A., and Bruce W. MacDonald, *Special Report: Emerging Disruptive Technologies and Their Potential Threat to Strategic Stability and National Security*, Federation of American Scientists, September 2018. As of March 26, 2020:
https://fas.org/wp-content/uploads/media/FAS-Emerging-Technologies-Report.pdf

Braun, Virginia, and Victoria Clarke, "Using Thematic Analysis in Psychology," *Qualitative Research in Psychology*, Vol. 3, No. 2, 2006, pp. 77–101.

Converse, Jean M., and Stanley Presser, *Survey Questions: Handcrafting the Standardized Questionnaire*, Quantitative Applications in the Social Sciences, No. 63, Sage Publications, 1986.

Hastie, Reid, and Robyn M. Dawes, *Rational Choice in an Uncertain World: The Psychology of Judgment and Decision Making*, Sage Publications, 2009.

Intel, "Intel Next 50 Study," August 22, 2018. As of March 26, 2020:
https://newsroom.intel.com/wp-content/uploads/sites/11/2018/08/intel-next-50
-study-results.pdf

Manyika, James, Michael Chui, Jacques Bughin, Richard Dobbs, Peter Bisson, and Alex Marrs, *Disruptive Technologies: Advances That Will Transform Life, Business, and the Global Economy*, McKinsey Global Institute, May 2013. As of March 26, 2020:
https://www.mckinsey.com/~/media/McKinsey/Business%20Functions/
McKinsey%20Digital/Our%20Insights/Disruptive%20technologies/MGI_
Disruptive_technologies_Full_report_May2013.ashx

Miller, David T., *Defense 2045: Assessing the Future Security Environment and Implications for Defense Policymakers*, Center for Strategic and International Studies, November 2015. As of March 26, 2020:
https://espas.secure.europarl.europa.eu/orbis/sites/default/files/generated/document/en/151106_Miller_Defense2045_Web.pdf

MIT Technology Review, "10 Breakthrough Technologies 2020," February 26, 2020. As of March 26, 2020:
https://www.technologyreview.com/lists/technologies/2020/

Morgan, M. Granger, *Theory and Practice in Policy Analysis*, Cambridge University Press, 2017.

Morgan, M. Granger, Baruch Fischhoff, Ann Bostrom, and Cynthia J. Atman, *Risk Communication: A Mental Models Approach*, Cambridge University Press, 2002.

NAE—*See* National Academy of Engineering.

National Academy of Engineering, "NAE Grand Challenges for Engineering," 2017. As of March 26, 2020:
http://www.engineeringchallenges.org/challenges/11574.aspx

Price, Vincent, and Peter Neijens, "Opinion Quality in Public Opinion Research," *International Journal of Public Opinion Research*, Vol. 9, No. 4, 1997, pp. 336–360.

Schuman, Howard, and Stanley Presser, *Questions and Answers in Attitude Surveys: Experiments on Question Form, Wording, and Context*, Sage Publications, 1996.

Slovic, Paul, Baruch Fischhoff, and Sarah Lichtenstein, "The Psychometric Study of Risk Perception," in V. T. Covello, J. Menkes, and J. Mumpower, eds., *Risk Evaluation and Management*, Springer, 1986, pp. 3–24.

Tourangeau, Roger, "Cognitive Sciences and Survey Methods," in *Cognitive Aspects of Survey Methodology: Building a Bridge Between Disciplines*, Vol. 15, National Academy Press, 1984, pp. 73–100.

WEF—*See* World Economic Forum.

World Economic Forum, "Top 10 Emerging Technologies 2019," June 2019. As of March 26, 2020:
http://www3.weforum.org/docs/WEF_Top_10_Emerging_Technologies_2019_Report.pdf

———, "The Global Risks Report 2020," *Insight Report*, 15th Ed., 2020. As of March 26, 2020:
https://www.weforum.org/reports/the-global-risks-report-2020